ARCHIMEDES
and the
ROMAN
IMAGINATION

ARCHIMEDES

and the

ROMAN
IMAGINATION

Mary Jaeger

The University of Michigan Press / Ann Arbor

Copyright © by the University of Michigan 2008
All rights reserved
Published in the United States of America by
The University of Michigan Press
Manufactured in the United States of America
⊚ Printed on acid-free paper

2011   2010   2009   2008   4   3   2   1

A CIP catalog record for this book is available from the British Library.

Library of Congress Cataloging-in-Publication Data

Jaeger, Mary, 1960–
     Archimedes and the roman imagination / Mary Jaeger.
          p. cm.
     Includes bibliographical references and index.
     ISBN-13: 978-0-472-11630-0 (cloth : alk. paper)
     ISBN-10: 0-472-11630-4 (cloth : alk. paper)
        1. Archimedes—Biography.   2. Mathematicians—Greece—Biography.
     3. Mathematics, Ancient.   I. Title.

     QA29.A7J34   2008
     510.92—dc22
     [B]   2007037351

For
Malcolm & Seth

PREFACE AND
ACKNOWLEDGMENTS

THIS PROJECT RISES from two sources. The first is a long-standing interest in teasing out what ancient literary descriptions of monuments tell us about the authors who describe them, figures whom I anachronistically imagine standing in front of each of these monuments and waving wildly at a camera, so as to draw attention from the monument to themselves. From this interest emerged the earliest-drafted chapter of this book, a reading of Cicero's account, in the *Tusculan Disputations*, of his discovery of Archimedes' tomb. That chapter, a version of which appeared in a 2002 volume of the *Journal of Roman Studies*, in turn generated two others, which discuss not monuments, strictly speaking, but artifacts: one explores Cicero's representation of Archimedes' spheres; the other examines Petrarch's use of Archimedes' story in his portrayal of his own relationship to books, especially the text of the *Tusculan Disputations*.

The second point of origin is a fascination—and frustration—of even longer standing, with one of the realities we face when studying figures from the ancient world, especially those about whom we know very little. That is the elusive nature of the biographical "fact," which, when released from narrative biographies and returned to its origins, loses its distinctive coloring as a biographical fact and fades into the underbrush as a component of the rhetoric of its sources. What follows is a modest attempt to track a few such facts to their dens in a consciously limited way, by choosing a historical individual for whom they are relatively few and then following them only insofar as they help answer some questions raised by my interest in monumentality: Why was Archimedes important to Cicero? How did their relationship extend backward and forward in time to include other relationships, both historical and literary? What can a close look at the figure of Archimedes tell us about the tradition that created it? According to a couplet attributed to the second-century BCE poet Licinius Porcius and viewed by scholars as possible evidence about the origins of Latin literature, the Muse came to Rome during the Second Punic War.[1] The flip side of the story is that Archimedes did not: he died during that same war, in a city his machines had defended against the Romans. Exploring the way in which tradition retells the stories of his life and death, with an eye to what was at stake for the sources that preserve them, can give us additional insight not only into the construction of what one scholar has called the "'true' and ancient Roman character" and "the 'Greek character' with which it was contrasted" but also into the greater narrative, now under extensive reconstruction, about Greco-Roman cultural interaction.[2]

I have taken a long time to write this book; a happy side effect has been the accumulation of many debts of gratitude over the years. It is a pleasure to thank the people who have provided resources, audiences, criticism, and advice. They include the co-organizers of a conference on memory, at Harvard in 1995 (the late Paige Baty, as well as Chris Bongie, Jennifer Fleischner, Dagmar Herzog, and Edward Wheatley, all working under the auspices of the Mellon Faculty Fellowship program and the Center for Literary and Cultural Studies). An invitation to a conference in Trier during the still-peaceful summer of 2001 allowed me to present work on Livy's account of the fall of Syracuse. I am grateful to the organizers and especially to Ulrich Eigler for his gracious hospitality. Audiences at meetings of the Classical Association of the Pacific Northwest merit special thanks for patiently listening to various dissections of the Archimedes story. I am grateful to the Classics Department at the University of Tennessee for a speaking invitation in the fall of 2004, which helped shape the chapter on the death of Archimedes, and to Elizabeth Sutherland for hospitality in Knoxville. Victoria University, Wellington, with its collegial faculty and excellent library, provided a cheerful and spectacular setting in which to draft the Petrarch chapter during a sabbatical in the spring of 2005; speaking invitations there and at Otago and Canterbury universities provided opportunities for more feedback on other parts of the book. I am especially grateful to Stephen Epstein and Mi-Young Kim for arranging the stay in Wellington and doing so much for me and my family while we were there and to John Davidson, Arthur Pomeroy, Babette Puce, David Rosenbloom, and Matthew Trundle for hospitality and conversation. William Dominik at Otago offered very helpful comments and criticisms; likewise Enrica Sciarrino and Patrick O'Sullivan at Canterbury. Gustavus Adolphus College invited me to speak on the "Eureka" story in the spring of 2006; I extend my thanks to the faculty there, especially Stewart Flory, Pat Freiert, and Will Freiert, three of the reasons why I went into Classics.

Andrew Feldherr, Thomas Habinek, Bill Keith, Chris Kraus, and Michele Lowrie read and commented on various early chapters; Sharon James, Matthew Roller, and my colleagues at Oregon, John Nicols, Ben Pascal, and Malcolm Wilson, read and gave advice on parts of the book in its later stages. I am grateful to them all. I also owe tremendous scholarly debts to several people whom I have not met. Their names appear in the text and notes, but I want to draw attention to them here, because their legwork made this book possible. They are Marshall Clagett, Reviel Netz, Antonio Quaquarelli, Christopher Rorres, and D. L. Simms.

I am grateful to Chris Hebert, the Classics editor for the University of Michigan Press, who chose readers both sympathetic and critical; and I thank those readers for identifying themselves, so that I might discuss aspects of the project with them. Andrew Feldherr's careful reading of the entire manuscript clarified the argument, caught mistakes, and helped give the book its final form; Reviel Netz both showed how this book could be much better and generously sent me electronic copies of two forthcoming works. (That I could meet his suggestions only part way is an indication of my scholarly limitations—they were fine suggestions.) Given all this help, it should not be, but this book still has its flaws and, inevitably, its errors. They are, all of them, my own.

Finally, I thank Malcolm Wilson for many things, including math advice and sanctuary on the Siuslaw River, and Seth Wilson for ongoing inspiration. This book is for them.

CONTENTS

ABBREVIATIONS

AJP   *American Journal of Philology*
ANRW   *Aufstieg und Niedergang der römischen Welt*
CA   *Classical Antiquity*
CAH   *Cambridge Ancient History*
CIL   *Corpus Inscriptionum Latinarum*
CJ   *Classical Journal*
CP   *Classical Philology*
CQ   *Classical Quarterly*
CW   *Classical World*
HRR   *Historicorum Romanorum Reliquiae*
HSCP   *Harvard Studies in Classical Philology*
JHS   *Journal of Hellenic Studies*
JRS   *Journal of Roman Studies*
LEC   *Les Études classiques*
LSJ   H. G. Lidell, R. Scott, H. S. Jones, *A Greek-English Lexicon*
MD   *Materiali e discussioni per l'analisi dei testi classici*
OED   *Oxford English Dictionary*
OLD   *Oxford Latin Dictionary*
PCPS   *Proceedings of the Cambridge Philological Society*
RE   *Real-Encyclopädie der klassischen Altertumswissenschaft*
REL   *Revue des études latines*
TAPA   *Transactions of the American Philological Association*
ZPE   *Zeitschrift für Papyrologie und Epigraphik*

INTRODUCTION:
THE "LIFE OF ARCHIMEDES"

IN THE FIRST CHAPTER of his dissertation on Archimedes, published in 1879 as *Quaestiones Archimedeae,* Danish scholar Johan Ludvig Heiberg methodically sets forth a narrative biography of the great mathematician.[1] The information about the lives of all the ancient mathematicians is meager and widely scattered, says Heiberg, but we know more about Archimedes than the others for two reasons: his defense of Syracuse, which drew the attention of historians; and his practice of prefacing his treatises with letters to his friends. Having identified these two strands of evidence, Heiberg constructs his narrative by bringing together material from both.

Before doing so, however, Heiberg points out that Eutocius's commentary on *The Measurement of the Circle* referred to an ancient biography of Archimedes by one Heracleides. This piece of information leads to a dead end, for we know nothing more about this biography or the biographer himself. Indeed, Heiberg has to justify calling him "Heracleides," because Eutocius calls him "Heracles" elsewhere.[2] This life of Archimedes (probably called simply Ἀρχιμήδους βίος, "Life of Archimedes") would have taken the shape imposed on it by its author's interests, prejudices, and understanding of how a "life" was supposed to be written. From our knowledge of Hellenistic biography, we can conjecture that it was probably more descriptive than narrative; that is, it probably focused on character more than event and included "anecdotes, witticisms, and eccentricities," precisely the kind of material, indeed, that the literary tradition has preserved about Archimedes, material familiar to many moderns in the form of the "Eureka" story and the statement "Give me a place to stand and I will move the world."[3]

Heiberg presses on, through Archimedes' possible date of birth in 287 BCE, his relationship to the Syracusan royal family (sources for the latter of which are Plutarch's *Life of Marcellus* and Archimedes' own letter to Gelon, which prefaces his *Sand-Reckoner*), his stay in Egypt, his amazing machines, his theoretical interests (symbolized by the sphere and cylinder on his tomb, which, Heiberg notes, was neglected and lost by the Syracusans and rediscovered by Cicero), his defense of Syracuse against Roman attack, his death during the sack of the city, and the grief of the Roman general Marcellus at learning of it. In both text and notes, Heiberg is careful to make clear that his narrative is a compilation of information from sources that vary in reliability: when giving Archimedes' date of birth, he adds, *si Tzetzae credimus* (if we believe Tzetzes); a parenthetical *ut videtur* (as it seems), qualifies his statement that Archimedes spent a long time in Egypt; and he pronounces entirely false any assertions that Archimedes traveled to Spain.[4] Heiberg's footnotes, most of them quoting his sources verbatim, are twice as long as the biography itself. In short, his life of Archimedes is a model of judicious reconstruction, careful documentation, and restraint in interpretation.[5]

More recent scholars take it for granted that readers will understand the constructed nature of this narrative, which, for convenience, I shall call the "Life of Archimedes."[6] In volume 1 of his *Works of Archimedes*, Reviel Netz does not even include a biography and says that perhaps one ought not be attempted.[7] Netz's comment makes good sense, for to all appearances, the legends comprising much of this "Life of Archimedes" have at best a tenuous relationship to the reality of Archimedes' life; and the modern historian of science has other, more rewarding, questions to ask.[8] Having noted the "radically different" strands of tradition about Archimedes, Netz brackets the anecdotes, looks instead to Archimedes' letters and treatises, and, eschewing narrative, demonstrates how the man's personality emerges from his work.[9] When he discusses ancient mathematicians in general, in *The Shaping of Deduction in Greek Mathematics* and in his essay "Greek Mathematicians: A Group Picture," Netz demonstrates what can be learned about their lives by taking a demographic approach to information gleaned from the treatises and the corpus of ancient commentary.[10] This approach allows Netz to make several claims about the field as a whole: he argues for the "catastrophic," as opposed to incremental, origins of Greek mathematics; the rarity of mathematicians; their generally privileged origins; and the position of mathematical inquiry at the intersection of literacy and orality, as well its position on the boundary between the theoretical and the material.

Heiberg would go on to produce the Teubner editions of Archimedes and to recover the text of Archimedes' *Method* from a palimpsest found in Constantinople; Netz is producing the first full translation of Archimedes' works into English and is part of the team now working on that palimpsest, which vanished early in the twentieth century, resurfaced at Christie's in 1998, was sold to an anonymous buyer, and now resides in the Walters Art Museum.[11] In Heiberg and Netz, then, we have two prominent historians of mathematics working over a century apart, neither of whom has biography as a central preoccupation, but both of whom acknowledge an obligation toward it, either by including one or explaining its absence. Some of their methodology is the same: Heiberg draws on known practices in Greek nomenclature when he argues that the name of Archimedes' ancient biographer was Heracleides; Netz does likewise when he argues that Archimedes' family was from the artisanal class.[12] Yet their approaches to biography produce widely disparate results: Heiberg carefully constructs a narrative following an outline about which even he is skeptical, while Netz avoids narrative entirely and, in its stead, produces a general portrait of mathematicians' lives, in which the image of Archimedes is that of one small face among several, an image, as it were, of lower resolution and thus hazier outline. In short, their lives of Archimedes have taken the shape imposed on them by their authors' interests, purposes, and understanding of how a "life" is supposed to be written.

Strikingly different from Heiberg's sober account and Netz's rigorous bracketing is the "Life of Archimedes" related by the mid-nineteenth-century polymath and biblioklept Guglielmo Libri. It is worth a brief digression, for, in contrast to Heiberg and Netz, Libri uses Archimedes' life as part of an overtly moral program. Libri introduces his *Histoire des sciences mathématiques en Italie* by observing that the lack of such a history has made foreigners think that only poetry and the arts could prosper in the land of Archimedes and Galileo.[13] Put off, he says, by the dryness of previous histories, which proceed from work to work without offering a glimpse of the men behind the science, he will weave those men's lives into his history. Including personalities is important, adds Libri, because a people's intellectual state is bound to its moral and political condition. Accordingly, he will adopt a new point of view and bring to the fore the men of science as well as their ideas.[14]

Libri prefaces his work with a history of mathematics in ancient times, because, he says, all phases of civilization are bound to one another. After surveying the scientific contributions of the Etruscans, he turns to the Romans, who he claims destroyed learning in acquiring their empire.[15] Their greatest crime, according to Libri, was the death of Archimedes.[16] Libri then outlines the great geometer's life and death: Archimedes was born in about the 467th year after the founding of Rome; he was related to King Hieron II, according to Plutarch, but Cicero's disdainful expression suggests that he was of less illustrious birth—one is shocked (*bien choqué*), notes Libri, to find that Cicero called one of the most extraordinary men who ever existed a "humble little fellow."[17] Archimedes' discoveries place him at the head of ancient geometers, Libri continues, before listing some of the mathematician's major theoretical treatises. It was, however, Archimedes' mechanical devices that won the admiration of the ages. After mentioning a passage from Diodorus Siculus on Archimedes' invention of a screw for raising the water of the Nile, Libri observes that since Diodorus refers elsewhere to an analogous machine for removing water from mines in Spain, one might believe Archimedes traveled to Spain as well as to Egypt. As corroborating evidence, Libri points to "the authority of other writers" (*l'autorité d'autres écrivains*) and directs the reader to a note at the end of the volume. Here, "the authority of other writers" resolves itself into a passage quoted from a notebook of Leonardo da Vinci, reporting that he found a story, in a history of Spain, of Archimedes using a grappling hook against ships during the war there against the English (!).[18] Libri next rejects the "Eureka" story and, having turned to the defense of Syracuse, pronounces it unlikely that Archimedes burned up Marcellus's fleet with mirrors, because the story "is not found in the most ancient authors."[19]

Libri's skepticism about the bath and the mirrors seems rather selective, not only because it stands in sharp contrast to his acceptance of the flimsy evidence for Archimedes' Spanish travel, but also because he singles out another dubious instance of Archimedean applied science as well worthy of attention: Archimedes' supervision of the building of a great ship for Hieron II. This story, says Libri, shows the great mathematician obliged "to lower himself so far as to direct the construction of a vessel where there was a chamber devoted to the shameful pleasures of the king." A writer more skeptical than Libri might have added that this story, too, is not "found in the most ancient authors," for it stems from a passage in the third-century Athenaeus's compendium of table talk, the *Deipnosophistai*.[20]

By "a chamber devoted to the shameful pleasures of the king," Libri refers to what Athenaeus's description calls an Aphrodision (Ἀφροδίσιον . . . τρίκλινον), a "chamber dedicated to Aphrodite, with three couches," mention of which follows a list of other luxurious appointments: a gymnasium and promenades, garden beds and bowers. The ship also had a library, a room for baths, staterooms, stalls for horses, and tanks for fish, none of which receives comment from Libri. Close readers of Athenaeus's description might also note that although Archimedes was the overseer (ἐπόπτης) of the project, Hieron also brought together shipwrights and other artisans and placed the Corinthian, Archias, in charge, as architect (5.206f); they might note that the passage cites specifically Archimedes' contributions to the design: he found a means of launching the half-built ship (5.207b); he equipped it with weapons (5.208c); and he was the inventor of the screw that removed water from the bilge (5.208f). Libri draws no attention whatsoever to these contributions. Having sniffed out the only whiff of sexuality that emanates anywhere from any story about Archimedes—and even this only indirectly, for the "shameful pleasures" are not Archimedes' but the king's—Libri is determined to make the most of it.[21]

For good reason! A fervent patriot and advocate of pure science, Libri wrote his history of mathematics with the lofty intent of providing moral examples: "In writing this history, my purpose has not been merely scientific. I also wanted to trace the life of illustrious scholars and depict the noble and generous spirit that led them to pursue (ceaselessly and despite thousands of dangers) instances of truth which could only be reached at the cost of sacrifices and unhappiness."[22] Libri, who, in his unrestrained patriotism, presents Archimedes as an Italian surpassing Greeks,[23] intended to restore the intellectual traditions of Italy and, by doing so, to save Italian youth from its self-destructive tendencies—hence his disdain for both applied science and sex. Yet Libri's reasons for pointing out the scandalous chamber of Aphrodite are not only moral. His "Life of Archimedes" reflects his pride in disclosing the obscure, overlooked, and otherwise forgotten. Expressing outrage at the chamber of Aphrodite allows Libri to point out that this important matter has gone unnoticed until now (*un fait qui mérite beaucoup plus d'attention, et qui a passé jusqu'à présent presque inaperçu . . .*), just as the suggestion that Archimedes traveled to Spain allows Libri to cite the passage from Leonardo's manuscript and, in doing so, to draw attention to his own discovery.[24]

I discussed the modern historians Heiberg, Netz, and Libri as biographers in order to illustrate some general truths about biographical writing: biographers present their subjects in ways that reflect their own interests and aims (Heiberg aims to demonstrate the mastery of sources required for the awarding of a doctorate—not his only goal, but one of them—while Libri aims to display his own discoveries even as he sustains the interest of his readers and improves them morally as well as intellectually); in addition, biographers must shape the events of a life into narrative, which means using the rhetorical tools of narrative; moreover, they rely for sources on materials originally meant to serve a variety of nonbiographical purposes. The "Life of Archimedes" is no exception. Its sources include histories, speeches, an epic poem, philosophical dialogues, collections of moral examples, a treatise on architecture, a compendium of table talk, a poem on weights, and a biography written about someone else. Anyone writing a narrative account of Archimedes' life must identify biographical material in this wide array of sources, extract that material (including anecdotes, witticisms, and eccentricities) from its immediate contexts, and then reassemble it into a narrative—that is, transfer it from one context to another.

To speak of "extracting material" is, of course, a metaphor. Extracting what I will generously call a biographical fact about Archimedes from a Ciceronian dialogue, for example, does not physically remove from the dialogue the words that express that fact.[25] Yet using the dialogue as a source for biographical information has an effect on readers of both biography and dialogue. On the one hand, readers of the biography have a highly restricted view of the text from which the information was taken. They see only the extracted fact or perhaps the immediate passage itself (if it is quoted, as in Heiberg's notes); in any case, they do not see the larger framework of the dialogue. On the other hand, once that same passage is known as the source of information valuable to an external narrative (in this case, the "Life of Archimedes"), it can become otherwise invisible to readers of the dialogue.[26]

Accordingly, the present book asks what insight we can gain and what different narrative we can shape when, having broken the "Life of Archimedes" into its constituent anecdotes, witticisms, and eccentricities, we return those elements to their contexts, reconsider them in their settings, and examine their relationship to those settings. The goal is to discover what the recontextualized material can tell us about the figure of Archimedes (not about the man) and what light that figure, used as a hermeneutic tool, can in turn reflect on both the works in which it appears and the fragments of the classical literary tradition that make up the greater context.[27] Netz uses context—cultural, social, and economic—to produce a group picture of ancient mathematicians. The following pages use context to attempt a "group picture" of a different sort: one man's portrait as created by a group, a portrait that ends up giving insight into those who contributed to it and their interrelations.

Where Netz's group picture bracketed the anecdotes, this group picture brackets the mathematical works. This book, then, is not about math or the history of math; nor does it attempt to ascertain the historicity of the traditions about Archimedes or the nature of some of his inventions—as D. L. Simms has done with such wit and verve in his many articles published in the journal *Technology and Culture* and elsewhere.[28] Instead, it offers a biography of the figure of Archimedes—a metabiography, as it were—that traces the career of that figure from the Hellenistic age through late antiquity and then concludes with a look at its rediscovery by the humanist Petrarch.[29]

Why Archimedes? After all, a close look at many an ancient figure can contribute to our understanding of the historical and literary contexts that produced it. From T. P. Wiseman's study of the traditions about Remus, for example, there has emerged the outline of a second and suppressed plebeian Rome, one shaped by the preliterary dramatic traditions of the fourth century.[30] Likewise, Judith Ginsburg's study of the traditions about the younger Agrippina has aligned Agrippina with both Augustus and Sejanus, thus exposing "the fallibilities of the regime."[31] Nevertheless, there are compelling reasons for students of classical—particularly Latin—literature to pay close attention to the figure of Archimedes, even if they are not historians of science.

First, Archimedes was the greatest mathematician the ancient world produced, yet few of the writers who make up the anecdotal tradition give evidence of having read his theoretical treatises. To be sure, Plutarch tells of the elegance of Archimedes' proofs, and Vitruvius writes of having gathered material on mechanics in the *commentarii* of Archimedes and various authors (indeed, Vitruvius may even have made a Latin anthology of such writings).[32] Yet Cicero, who invokes the figure of Archimedes more frequently than any other classical author, refers to none of his treatises;[33] nor do Livy, Pliny, Quintilian, Valerius Maximus, or Silius Italicus. Archimedes' fate appears to have anticipated that of Einstein, a symbol of intellect known and celebrated by many (through T-shirts, posters, and mugs), but a thinker read and understood by very few.[34] The figure of Archimedes meant something different to the rhetoricians and poets than the mathematician himself did to his correspondents. Study of this figure can tell us something more about how the Romans presented themselves as thinkers.

Second, there is the particular historical context of Archimedes' death, an event that retrospectively shaped many of the stories about him. The sources generally agree that Archimedes died during the Roman sack of Syracuse in 212/211 BCE, at the hands of a Roman soldier (see chap. 4). Although we do not know how old he was at the time of his death (the Byzantine poet Tzetzes, who says he was seventy-five, probably chose an aesthetically pleasing number of years), the responsible historian Polybius emphasizes that it was an elderly Archimedes who defended Syracuse.[35] We can take as given, then, that Archimedes' adult life overlapped with a long period of significant interaction between Syracuse and Rome. Rome besieged Syracuse early in the First Punic War (during 263 BCE, in a war that lasted from 264 to 241 BCE) but then formed an alliance with Hieron II that lasted until his death in 215 BCE. The First Punic War saw extensive fighting in and around Sicily; and after their eventual victory over Carthage, the Romans first acquired western Sicily as a province and then, after the Second Punic War, gained the rest of the island.[36] At the same time, the Second Punic War seems to have resulted in a growing contempt on the part of Rome for Sicily and Magna Graecia and in increasing interest in "real" Greece.[37] In consequence, the figure of Archimedes, defender of Syracuse and victim of Roman violence, provides a focal point for ideas and questions about the good and bad effects of Greco-Roman interaction, Roman expansion, and the relative merits of the Sicilian, Greek, and Roman national characters.

Third, the events of the first and second Punic wars provide the historical background for the creation of early Latin literature, which was invented by writers of varied origin responding to Greek models.[38] Livius Andronicus, a Greek from Tarentum captured during the war with Pyrrhus, is believed to have produced the first real Latin play, in 241 BCE, by adapting a Greek original. He also translated the *Odyssey* into Saturnians, a native Italic meter. Gnaeus Naevius, probably from Campania, fought in and composed Rome's first national epic, the *Bellum Punicum,* on the First Punic War. The earliest surviving written versions of Roman comedies, based on Greek forms, emerge from this period, which almost brings us down to the performance dates of the comedian Plautus's earliest plays. Fabius Pictor, who took an active part in the events of the Second Punic War, produced the first Roman history by a Roman, although he wrote it in Greek. He may have brought his narrative down to the end of the war.[39] Cato the Elder, author of perhaps the first real Roman history in Latin, was a military tribune in Sicily in 214 BCE; the fifth book of his *Origines* covered the Second Punic War. Quintus Ennius, who became the leading poet of the next generation and produced the first Roman national epic in Latin hexameters, served in Sardinia during the Second Punic War. Although none of the works or fragments of works surviving from these pioneers of Roman literature mentions Archimedes, he lived his life among the same events that inspired and informed these writers. Thus many of the anecdotes that make up the "Life of Archimedes" took shape during the extraordinarily creative last half of the third century BCE, as the Roman system of commemorating aristocratic achievement became more coherent and complex, within the cultural context of an emerging Latin literature and an increasingly literate and literary debate about Roman cultural identity.[40]

Given these historical circumstances, it is not surprising that many of the stories of Archimedes' life come from authors self-consciously straddling the Greco-Roman cultural divide. Polybius wrote for a Greek-reading audience—which would have included upper-class, educated Romans—and explained to it how and under what form of government the Romans came to dominate the known world in a mere fifty-three years. Cicero "translated" Greek philosophy into Latin—thus creating a Latin philosophical language and tradition for Rome—and ostentatiously placed his philosophical works in the context of the Roman appropriation of Greek culture. Plutarch set Greek lives against Roman ones. We have these stories about events at Syracuse during the Second Punic War because they were of use to writers taking a Roman point of view or explaining Rome's expansion and culture to both Greeks and Romans. To writers taking this cross-cultural approach, Archimedes, a figure emblematic of the stereotypically Hellenic obsession with the abstract, becomes "good to think with." The following pages examine generally how these writers and others represent this life story from the period of Rome's literary beginnings and its great struggle against Carthage, and specifically how they fit its elements into their own arguments about Roman cultural achievement.

Syracuse itself lends richness to the context. Wealthy and powerful in fact, the city was also famous in the literary tradition.[41] Thucydides' use of the place is the most relevant to the present inquiry. Books 6 and 7 of his history of the Peloponnesian War told of the Athenian expedition against Sicily (415–413 BCE), its failure to capture Syracuse, the Athenians' defeat and failed retreat, and the wretched fates of those taken prisoner.[42] Thucydides' grim narrative and detailed representation of the landscape in which the siege, the retreat, and the deaths of the prisoners took place left to the historiographical tradition a memorable Syracuse, both as setting to reuse and as a literary context to invoke. The writers who described the Roman attack two hundred years later (214–212/211 BCE)—Polybius, Livy, and eventually Silius Italicus—were interested in significant historical repetition and willing to draw on this distinguished historical tradition.[43] They could expect readers to remember the climactic events of what Thucydides had argued was the greatest disturbance that the world had known and to envision events at Syracuse during the Second Punic War against the backdrop of the Sicilian Expedition. The stories of Archimedes' life thus take on ideological overtones from the city's literary past, and those overtones contribute to later writers' responses to contemporary events and their attempts to shape them.

The figure of Archimedes that emerges from these sources is, in general, one generated by Roman reaction and regard. For the most part, then, this study explores the Roman use and reuse of Archimedes' story. As a corollary, it examines the way in which various authors present themselves as participating in the story's reception and transmission. The "Life of Archimedes" lends itself nicely to the study of the complexities of reception by audiences both within and external to a text or set of texts, because so many of the anecdotes themselves revolve around ideas of reception and response as they relate the reaction of various audiences to Archimedes' words, his actions, and the events of his life. In response to Archimedes' claim that if there were another world, he could go to it and move this one, Hieron asks him to prove it; when Archimedes demonstrates it by launching the big ship, Hieron responds by asking Archimedes (in Plutarch) to design the defenses of Syracuse and by declaring (in Pappus) that, henceforth, anything that Archimedes says must be believed. Marcellus sees Archimedes' machines dash Roman warships against the walls of Syracuse and responds with a joke; he hears of Archimedes' death and responds to it with various degrees of grief, anger, and attempts at compensation. As audiences receive these stories and pass them on, the information accompanying them becomes part of the stories, telling how they are received and passed on not only by characters within the text, such as Hieron and Marcellus, but also by outside characters (including authors and the unnamed subjects of, e.g., the verb *ferunt*, "they say"). The process of reception culminates in the autobiographical passages of Cicero and later Petrarch, each of whom uses the figure of Archimedes to help represent his own rediscovery and transmission of the past.[44] This book tries to convey how the "Life of Archimedes" changes meaning as it is interpreted by members of various reading communities, not the modern scholarly community or the mathematicians and modern scholars of history and literature whose "lives" I have surveyed here, but communities composed of elite Romans and Greeks of the republican era, nonelite Romans of the Augustan Age, imperial Romans and Greeks, and Renaissance readers.[45]

Part of the process of considering context and reception is to consider how, precisely, the stories about Archimedes are embedded within these contexts. Examining the anecdotes, I ask narratological questions: Who speaks? When does Archimedes himself speak? When do narrators themselves speak, and when do they convey information through indirect speech? How does a particular mode of speech representation affect the story told by redirecting attention and responsibility? My approach to evidence in the forms of narrative and the embedding of anecdotal evidence into narrative has been generally influenced by the work on narratology done by Gérard Genette and refined by Mieke Bal.[46] A study that examines multiple accounts of the same event cannot but make use of the difference between *histoire, récit,* and *narration,* and my approach draws implicitly on Genette's categories of narrative analysis (especially those of narrative mood and voice).[47] I have, however, avoided as much as possible using the technical terms of narratology, partly to prevent correct labeling from becoming an end in it itself and partly to make the discussion easily accessible to the general reader.[48]

Some of my discussion responds specifically to Andrew Laird's ideas about speech presentation, especially as he sets them forth in the first two chapters of his book *Powers of Expression, Expressions of Power.* Laird lucidly brings questions of ideology to bear on the formalist study of narrative.

> No text or discourse will ever represent a state of affairs neutrally. However comprehensive or disinterested a writer or speaker sets out to be, the inevitable determinations of omissions, closure, provenance, and genre will mean that the account produced will be angled, selective, partial and ideological. (16–17)

The "Life of Archimedes" is a case in point. No anecdote about Archimedes comes from the pen of an eyewitness or even from the pen of a writer citing an eyewitness. Our earliest anecdotal evidence for Archimedes' life appears in the work of Polybius, who wrote a generation and a half after the end of the Second Punic War. Polybius describes Archimedes' success at defending Syracuse from the Romans, but the part of the history that told of the city's fall and might have included the story of Archimedes' death is lost. Polybius is a particularly valuable source, because he could have spoken to eyewitnesses.[49] Yet even an eyewitness account will be colored by the interests and preconceptions of the eyewitness. The next surviving reference to Archimedes is made by Cicero, who served as quaestor in Sicily in 75 BCE and mentions Archimedes' death in a speech he made in 70 BCE when prosecuting Verres, former governor of Sicily (*Verr.* 2.4.131). Cicero's later speeches, letters, and philosophical works mention Archimedes' devotion to study, his spheres, and his tomb.[50] To paraphrase Laird, none of these stories point to a contemporary "control" version from which the others deviate.[51] Moreover, the writers who told stories about Archimedes, many of them Romans, do not even set out to be disinterested. The stories told and retold by Romans about a Sicilian who delayed the

Roman capture of Syracuse for three years and was killed by a Roman cannot but be "angled, selective, partial and ideological." Part of the present project is to show how this skewing came about and what it aims at; accordingly, this book offers a case study in embedding which will, I hope, contribute incrementally to our understanding of how ideology and partisan politics affect the way in which authors present biographical material and rhetorical exempla.

This books falls into three parts, which deal, first, with the creation of Archimedes as a figure embodying ideas of invention and transmission; second, with that figure's role in narratives of Greco-Roman interaction; and third, with its role in narratives of the loss and recovery of intellectual traditions. Returning these stories to their contexts and examining their rhetorical function within those contexts requires us to look at them with new eyes. Chapter 1, more explicitly than the others, is an exercise in such defamiliarization. It reconsiders the most famous story about Archimedes, which tells of him leaping from his bath with the cry εὕρηκα! The core of the story, the leap from the bath and the cry, has become so familiar and been abstracted from its context to such a degree that some of the anecdote's most interesting details have been left behind.[52] Chapter 1, accordingly, reconsiders the story within its immediate context first, then in its broader one. The result of this defamiliarization is to show both how the figure of Archimedes becomes a topos for discovery and how the idea of reception participates in shaping the story of discovery.

Cicero, who uses the figure of Archimedes in his speeches, letters, and philosophical dialogues, appears to be the most influential surviving filter through which elements of Archimedes' biography have passed.[53] Chapters 2 and 3 examine closely some of the Archimedes stories as they appear in Cicero's texts, to determine precisely how Cicero used them and thus to ascertain why the figure of Archimedes was so important to Cicero. Chapter 2 presents a close reading of Cicero's account in the *Tusculan Disputations* of his discovery of Archimedes' neglected tomb and shows how the figure of Archimedes becomes emblematic of the dialogue's central ideas. Moreover, the narrative of reception within Archimedes' story, as Cicero recalls what he knows about Archimedes and the tomb, is transformed into a narrative of reception external to Archimedes' story, as Cicero becomes a participant in it by finding the tomb and writing about his discovery. The rediscovery of Archimedes' tomb thus becomes emblematic of Cicero's "rediscovery"—his appropriation—of Greek learning for Rome.

Cicero's *De republica, De natura deorum,* and *Tusculan Disputations* preserve the earliest descriptions of a sphere, or orrery, attributed to Archimedes, and later appearances of this sphere appear to rely on those descriptions. Chapter 3 of the present study examines the sphere passage in the *De republica* and asks why it involves not one but two spheres. A close reading of this passage, which returns it to its context within the material framing the dialogue, shows how the comparison of the two spheres participates in the generation of dialogue and thus acts as an extended metaphor for the transfer of Greek cultural capital to Rome. Roman, Sicilian, and Greek alike benefit from this transfer, because dialogue is a process in which eventually all can engage. For my discussions in chapters 2 and 3 especially, I owe a great debt to Thomas Habinek and the instrumentalist view of Latin literature conveyed by his book *The Politics of Latin Literature* (particularly chap. 2, "Why Was Latin Literature Invented?").[54] The coda that concludes the first part of the present study considers the reception and rereading of Cicero's sphere passage in the fourth-century astrological treatise of Julius Firmicus Maternus. It argues that later writers who use the sphere imitate Cicero's manner of using it rather than the description of the sphere itself.

The argument introduced in part 1 runs through the rest of the book: the figure of Archimedes was important to many later writers because it was important to Cicero, and its appearance in Cicero shapes the manner in which they use it. Part 2 argues a related point: Archimedes was important and useful to Cicero mostly because he was important to Marcellus, the Roman conqueror of Syracuse. Exploring a wide variety of sources for the siege and sack of Syracuse, part 2 examines the figure of Archimedes as it relates to that of Marcellus. Chapter 4 examines the various accounts of Archimedes' death. My general narratological approach to this part of the "Life of Archimedes" has been influenced by an observation made by Laird when he discussed Socrates' retelling of part of the *Iliad*. First, Laird pointed out that Socrates makes only two changes: he uses prose instead of verse and indirect, instead of direct, speech. Laird then said, "a change of narrative—even on a stylistic level—entails a change of what that story *means* to its audience—not least where its ethical effects are concerned."[55] Read with an eye to "what the story *means* to its audience," especially its ethical effects, the account of Archimedes' death, told by writers using different narrative styles, makes it evident that the event was a political problem for Marcellus, one that received considerable spin in the generations after the war.

Chapter 5 examines the defense of Syracuse, with particular attention to Plutarch's version of the conflict and the role played by Archimedes in delineating Marcellus's character. An examination of Marcellus's response to the absurdity of the situation in which he finds himself and his ships at Syracuse forms the center of an analysis of the ways in which Plutarch uses the two men's images to shape each other. In doing so, he indicates limits and boundaries: the limit of Roman Hellenization, the limit of Roman planning and even courage, the impassable boundary between his exemplary Roman and Platonic ideal of a Greek. The coda to part 2 shows how recognizing the ideas of absurdity inherent in the account of the defense of Syracuse enhances our reading of another text from late antiquity, Claudian's short poem on Archimedes' sphere.

Following one thread of argument, that of discovery and transmission, the last part of this book turns to the humanist Petrarch and his "rediscovery" of Archimedes. Readings of passages from two of Petrarch's Latin prose works, the *De viris illustribus* and the *Rerum memorandarum libri*, show how his "Life of Archimedes" reflects the tension between his attempts to recover classical texts and his Christian and Neoplatonist belief in the inevitable decay of the material world.

Each narrative element of the "Life of Archimedes" has suggested its own line of approach. Consequently, each of the following chapters was conceived separately and can be read as an independent case study interpreting a given anecdote, text, or set of texts. Nevertheless, after a great deal of thought and experimental shifting around of material, I have arranged the chapters so that the first three take up the sources in reverse chronological order.[56] This is for the sake of the overall argument, which thus moves from the more familiar story to the less familiar and from the simpler example to the more complex. The result for the entire book has been that after tearing apart one narrative, I have succumbed to temptation and constructed another. This book thus presents what we might call an "Afterlife of Archimedes," a story of discovery and rediscovery, appropriation and transmission, loss and recovery.

PART ONE

CHAPTER 1

THE "EUREKA" STORY

Then Mr. Beebe consented to run—a memorable sight.

E. M. FORSTER, *A ROOM WITH A VIEW*

DURING MOST OF ARCHIMEDES' LIFE, Syracuse was ruled by Hieron II, who had come to power through a military coup and thus reigned as a tyrant—a usurping monarch—for some years before being proclaimed king. The story of Archimedes' helping Hieron solve a problem with regard to his authority, whether the events it relates happened or not, may reflect the tension of the early part of Hieron's reign, when he was securing his hold on the city and was perhaps more vulnerable to insult than he would have been later. Hieron had contracted to have a gold crown made as an offering to the gods. When the crown had been made and dedicated, an informant told him that the artisan who made the crown had substituted silver for part of the gold. This was an affront to Hieron's authority and a threat to his prestige, as well as an act of impiety. The problem was, how could one detect the crime without harming the crown? What had been dedicated could not be mutilated. (Indeed, as Hannibal the Carthaginian learned, taking a core sample from a votive offering for this purpose could invoke divine wrath.)[1] Hieron assigned the problem to Archimedes. Mulling it over as he sank into a bath, Archimedes noted that the more his body was submerged, the more the water overflowed the bath. The resulting insight—that the volumes of body submerged and water overflowing were related—made him leap from the bath for joy and run off shouting, "Eureka! Eureka!" He had discovered the principle of specific gravity and thus how to solve the problem of the crown.[2]

An English word in regular use, *eureka* functions allusively: it refers indirectly to the specific situation in which the first-person perfect active indicative of the verb εὑρίσκειν became something more than simply the Greek way of saying, "I have found"; that is, it evokes the memory of Archimedes leaping from his bath and running through the streets of Syracuse.[3] Yet the alluding *eureka* and the story of Archimedes' discovery are so familiar as to be taken for granted, and only when we turn to the sources (or to the rare biography that quotes scrupulously) do we see that *eureka* is not exactly what Archimedes said. In fact, he said, εὕρηκα, εὕρηκα—"again and again" (*identidem*), according to one source; "over and over" (πολλάκις), according to another.[4] Noted by the ancients, but generally absent from the English usage, this repetition was clearly an important feature of the original anecdote. I will return to this point, but for now I will focus on what else comes into view when we defamiliarize the story by returning it to its settings.

We do not know who first told the "Eureka" story, who first told it in Latin, or who first wrote it down. It surfaces early in the Augustan Age in Vitruvius's *De architectura*, then resurfaces roughly a century later in an essay from Plutarch's *Moralia*.[5] A version of the problem of the crown, without reference to the bath or the word *eureka*, appears in Remius Favinus's *Carmen de ponderibus*, a hexameter poem on weights and measures from the end of the fifth century CE, and a brief mention of it, also without bath or exclamation, appears in Proclus's *Commentary on the First Book of Euclid's "Elements,"* which is roughly contemporary with Favinus's poem.[6] I shall take up each of these sources in turn.

Vitruvius tells the "Eureka" story in the preface to his ninth book (on sundials), when he lists famous examples of mathematical "discovery" (*inventio*). Having described Plato's method for doubling the area of a square and Pythagoras's discovery of the theorem that bears his name, Vitruvius first acknowledges Archimedes' many different contributions, then says that the one he is about to describe seems to have been worked out with "cleverness that was unlimited" (*infinita sollertia*). Vitruvius tells how Hieron wanted to dedicate the crown; how he gave the goldsmith a certain amount of gold; how, after the crown was delivered, the allegation was made that silver had been substituted for some of the gold; and how Hieron—offended at having been taken for a fool, "but not finding by what method he might detect this theft" (*neque inveniens qua ratione id furtum reprehenderet*)—turned the problem over to Archimedes. This part of the anecdote concludes with Archimedes' moment of insight and his emotional response to it.

Tunc is, cum haberet eius rei curam, casu venit in balineam, ibique cum in solium descenderet, animadvertit, quantum corporis sui in eo insideret, tantum aquae extra solium effluere. Itaque cum eius rei rationem explicationis ostendisset, non est moratus, sed exiluit gaudio motus de solio et nudus vadens domum verius significabat clara voce invenisse, quod quaereret; nam currens identidem graece clamabat εὕρηκα, εὕρηκα. (9 pr. 10)

[Then, while he had this matter in mind, he came by chance to the baths and, as he lowered himself into the bathtub there, noticed that however much of his own body settled into the tub, that much water flowed out of it. And so, since he had revealed the method of explaining the matter, he did not delay but, stirred by joy, sprang from the tub and, going home-ward naked, kept indicating rather truly in a loud voice that he had found what he was seeking. For, as he ran, he called out again and again in Greek, "I have found! I have found!"]

Vitruvius goes on to describe how Archimedes "is said" (*dicitur*) to have pro-ceeded "from this commencement of discovery" (*ex eo inventionis ingressu*) to the proof. I will return both to the reference to tradition suggested by *dic-itur* and to the details of the leap from the bath. For now, let us continue reviewing the anecdote.

Vitruvius describes Archimedes' method of proof by telling it as a story. Archimedes filled a vessel to the brim with water, lowered a lump of silver into it, and measured the amount of water displaced. He did the same to an equal weight of gold, which displaced less water. Finally, lowering the crown into the vessel, he found that it displaced more water than did the lump of pure gold yet less than did the lump of silver. Having made some calcula-tions, Archimedes "detected" (*reprehendit*) the amount of silver in the crown, and "the contractor's theft was revealed" (*manifestum furtum redemptoris*).

To recast Archimedes as Sherlock Holmes, notes Jean Soubiran, is per-haps to treat the great mathematician with less dignity than he deserves.[7] Yet this story not only reflects the practical cast that Vitruvius gives many of his exempla but also is rhetorically effective in this particular context, more so than even the anecdote that one might expect in the preface to a book on sundials: an account of Archimedes' inventing the orrery, a sphere that showed the motions of the sun, moon, and planets.[8] In fact, the "Eureka" story helps Vitruvius make the argument, dear to heart of the modern academic and a commonplace in antiquity, that intellectual excellence, specifically the excellence of writers, or *scriptores,* deserves as much reward as athletic prowess.[9]

Vitruvius counts a wide range of intellectuals, poets, rhetoricians, philosophers, and mathematicians, including Archimedes, among these *scriptores.* Continuing on from Archimedes' proof, he reports Archytas's doubling of the cube by finding mean proportionals and Eratosthenes' doing the same by using the mesolabe, and he praises Democritus's work on physics. Vitruvius then adds that Latin authors (Ennius, Accius, Lucretius, Cicero, and Varro) have also made immortal contributions and that his ninth book will rely on all these authorities, both Latin and Greek (9 pr. 18): "And thus, Caesar, relying on these authorities [*auctoribus*] and applying their sensibilities and advice, I have composed [*conscripsi*] these volumes."[10] With the verb *conscripsi,* Vitruvius inserts himself into the company of *scriptores,* a varied group, whose membership extends from mathematicians to poets and from long-dead Greeks to recently dead Romans.[11] Thus it makes sense to reconsider the "Eureka" story both for what it contributes to Vitruvius's more specific argument about the relative values of intellectual and physical accomplishments and for what it contributes more generally to his own staking out of his position as a writer addressing Augustus.

The preface opens and closes by comparing athletes and intellectuals.[12] It begins with a reference to the victors at the Olympian, Isthmian, and Nemian games. The prizes awarded these men are so great that, says Vitruvius, "not only do they carry off praise while standing in the assembly with palm and crown [*cum palma et corona*], but also, when returning victorious to their own cities, they ride into their fatherland triumphing in a quadriga and enjoy an established tribute from the commonwealth all their lives long [*perpetua vita*]."[13] Vitruvius is amazed that writers do not receive prizes equal to and even greater than these, for athletes only strengthen their own bodies by their exertions (*quod athletae sua corpora exercitationibus efficiunt fortiora*), and their distinction (*nobilitas*) lasts only as long they live, whereas writers improve the minds of other people and can continue to do so over generations.[14] Writers, says Vitruvius, "offer advantages that are unlimited, forever, to all peoples" (*infinitas utilitates aevo perpetuo omnibus gentibus praestant*). The preface's opening sentences, then, set up a series of antitheses: athletes versus writers; bodies versus minds; the limited, temporary, and individual versus the unlimited, permanent, and universal. These themes resurface near the end of the preface (15), where Vitruvius points out that the ideas of the men he has mentioned "are gotten up for everyone's advantage forever" (*ad omnium util-itatem perpetuo sunt praeparata*), while "athletes' distinctions, however, soon grow old, together with their bodies" (*athletarum autem nobilitates brevi spatio cum suis corporibus senescunt*).[15] Moreover, by elaborately paraphrasing the opening of Isocrates' *Panegyricus,* Vitruvius conveys implicitly another set of antitheses: Greek original versus Roman rewrite (possibly as a counterpart to Isocrates' antithesis of Greek versus barbarian) and thinker versus writer. Whereas Isocrates speaks both of those who "labored privately on behalf of the common good and so equipped their own souls [τὰς αὐτῶν ψυχὰς οὕτω παρασκευάσασιν] as to be able to help others" and of the benefit to all of a single man who "thought well" (εὖ φρονήσαντος), Vitruvius characterizes intellectuals as writers (*scriptores*).[16] Vitruvius's emphasis, then, is on the concrete result of thought, the written and published work.

This set of antitheses helps to make sense of two details in Vitruvius's vivid and memorable image of Archimedes springing from the bath and making his way through Syracuse. Vitruvius says explicitly that Archimedes was naked (*nudus*), and having already said that Archimedes was making his way home (*vadens domum*), Vitruvius specifically identifies his gait: Archimedes was running (*currens*) as he shouted. These are details so obvious as first to go unnoticed and then, when we do notice them, to appear redundant or gratuitous.[17] Yet when we restore the "Eureka" story to its context of the opposition between the tribute paid to athletes as opposed to writers, these two details—Archimedes' nudity and gait—look less redundant and more pointed and emphatic: together with his bath, they assimilate him to his antithesis, the Greek athlete.

Fikret Yegül has observed that "the reciprocal (and sometimes antithetical) relationship between the gymnasium and the bath, culminating in the fusion of these two institutions by the end of the first century B.C., irrevocably linked exercise with bathing and baths with gymnasia in Roman culture."[18] The presence of palaestrae in the baths, as well as the paintings of nude athletes on their walls, also link the two activities.[19] Vitruvius himself brings together bathing and exercise in his description of plans for palaestrae (5.11.1–4), a passage that, although it does not follow the sequence of activity at the baths (exercise, then bathing), does anticipate the sequence of events in the "Eureka" story: it moves from the rooms for bathing, to the running track, to the paths that are to be at edges of the running tracks, so that people who are dressed (*vestiti*) and strolling are not impeded by exercisers.[20]

If Vitruvius's references to Archimedes' nudity and gait, taken together in the context of the preface, assimilate him to the Greek athlete, what are the effects of this assimilation? First, it makes the antithesis between athletic and intellectual achievement more pointed by drawing attention to the meaningful differences between them: the athlete's achievement is self-improvement limited to his body and his lifetime; Archimedes' cleverness, in contrast, was unlimited (*infinita*), like the "unlimited usefulness" of the contributions of writers (*infinitae utilitates*); whereas athletes are primarily concerned with their own bodies as ends in themselves, Archimedes pointedly was not.[21] Vitruvius has reworked the idea of the athlete's naked competition into an analogy: Archimedes is to the mind as victorious athletes are to the body. As a preliminary conclusion, note that by restoring the anecdote to its context, we have recovered the figure of Archimedes as intellectual athlete and have thus made it possible to see one way in which Vitruvius tailors it to support his argument. Most important, as we shall see, despite Archimedes' apparent lack of concern about his body, these details about his body and its movement make that body central to the story.

For Plutarch, the story of Archimedes and his bath is one of mind over body. A bathing Archimedes appears three times in Plutarch's works: in the *Life of Marcellus* 17 (see chap. 5 in the present study), in a discourse from the *Moralia* on the role of old men in public affairs (786c), and in a dialogue from the *Moralia* arguing that one cannot live happily following the precepts of Epicurus. This last is where we find Plutarch's version of the "Eureka" story.

Plutarch argues that Epicureanism denies its followers the pleasures of the mind, particularly those stemming from mathematics (1093D). Those pleasures are considerable, for Plutarch claims that geometry, astronomy, and harmonics have the potency of love charms and that Eudoxus, Aristarchus, and Archimedes share the ability to make a sort of music. After singling out the painter Nicias, who so loved his work that he had to ask his servants whether he had eaten, Plutarch says that even greater pleasure came to the geometers Euclid, Philip, Archimedes, Apollonius, and Aristarchus. He points out that although the love of pleasure assumes many forms, no one has ever been so overjoyed at having his way with a woman that he sacrificed an ox, nor has anyone prayed to die immediately if only he could eat his fill of cakes. In contrast, Eudoxus prayed to be consumed in flames if only he could stand next to the sun and ascertain the size, shape, and composition of the planets, while Pythagoras sacrificed an ox in honor of the occasion when he discovered his theorem. Then, like Vitruvius, Plutarch passes from Pythagoras to Archimedes.

Ἀρχιμήδη δὲ βίᾳ τῶν διαγραμμάτων ἀποσπῶντες συνήλειφον οἱ θεράποντες· ὁ δὲ ἐπὶ τῆς κοιλίας ἔγραφε τὰ σχήματα τῇ στλεγγίδι, καὶ λουόμενος ὥς φασιν ἐκ τῆς ὑπερχύσεως ἐννοήσας τὴν τοῦ στεφάνου μέτρησιν οἷον ἔκ τινος κατοχῆς ἢ ἐπιπνοίας ἐξήλατο βοῶν 'εὕρηκα' καὶ τοῦτο πολλάκις φθεγγόμενος ἐβάδιζεν. οὐδενὸς δὲ ἀκηκόαμεν οὔτε γαστριμάργου περιπαθῶς οὕτω 'βέβρωκα' βοῶντος οὔτε ἐρωτικοῦ 'πεφίληκα,' μυριάκις ἀκολάστων γεγονόντων καὶ ὄντων.

[*His servants used to drag Archimedes away from his diagrams by force to give him his rubbing down with oil; and as they rubbed him he used to draw the figures on his belly with the scraper; and at the bath, as they say, when he discovered from the overflow how to measure the crown, as if possessed or inspired, he leapt out shouting "I have found" and went off saying this over and over. But no one have we heard shout with similar rapture, neither a glutton "I have eaten," nor a gallant "I have kissed," though sensualists unnumbered have existed in the past and are with us now.*][22]

Like Vitruvius, Plutarch sets Archimedes' mind against his own body. (Archimedes uses his belly as a writing surface!) He also sets Archimedes against people concerned only with the body—not only his slaves, but also the glutton and the gallant—just as Vitruvius sets him against the athlete. Vitruvius made Archimedes into a comic parody of the athlete; now Plutarch offers up glutton and gallant as parodies of the intellectual. While Vitruvius mentions Archimedes' joy as the force ejecting him from the bath, Plutarch places emphasis throughout the passage on the importance of intellectual pleasure—no surprise in an essay arguing that following the tenets of Epicurus brings no real or lasting happiness.

Although Favinus and Proclus mention only the crown problem and not the bath, their versions share some noteworthy features with the others. The last third of Favinus's two-hundred-line *Carmen de ponderibus* tells how to demonstrate the fraudulent debasement of metals by using their specific gravity, and it credits Archimedes with the discovery.

Argentum fulvo si quis permisceat auro,
Quantum id sit quove hoc possis deprendere pacto,
Prima Syracosii mens prodidit alta magistri.
Regem namque ferunt Siculum quam voverat olim
Caelicolum regi ex auro statuisse coronam,

Conpertoque dehinc furto—nam parte retenta
Argenti tantundem opifex inmiscuit auro—
Orasse ingenium civis, qui mente sagaci,
Quis modus argenti fulvo latitaret in auro,
Repperit inlaeso quod dis erat ante dicatum.   (125–34)

[Should someone mix silver with tawny gold,
how much of it there is, and how you can detect it,
the deep mind of the Syracusan teacher was the first to disclose.
For they say that the Sicilian king had dedicated,
to the king of the gods, a crown made from gold, which he had once
* vowed,*

and that, when a theft was then discovered (for having kept
part of the gold, the maker blended in an equal weight of silver),
the king invoked the genius of his citizen, who, with wise intelligence,
* discovered*
what measure of silver was hiding in the tawny gold,
without harming the item that had been already dedicated to the gods.]

The poem goes on to tell readers precisely how they, too, can do this. The method it relates, as Reviel Netz has pointed out, draws on the findings that Archimedes set out in his treatise *On Floating Bodies*. A body submersed in water is subject to buoyancy equal in magnitude to the weight of the water that it displaces.[23] If one puts on the pans of a balance two objects having equal weight but different density—a crown of solid gold and a crown of gold-plated silver, for example—then places those pans in columns of water, the denser will sink further because it has less volume, displaces less water, and has, consequently, less buoyancy.

Although this method of discovery differs from the one presented by Vitruvius, Favinus's manner of introducing the story is similar. Archimedes is again the detective uncovering crime—and so, potentially, is the reader (*quove hoc possis deprendere pacto*). Once again, Archimedes takes on the problem at Hieron's request, and telling the story involves invoking tradition: Favinus introduces the story with "they say" (*ferunt*). The striking difference is that Favinus does not describe the method by telling a story of Archimedes' carrying out an experiment, as Vitruvius did; rather, in didactic mode, he addresses a series of instructions to the reader.[24]

Proclus, too, invokes tradition, when he argues for the utility of geometry.

Recall what Hieron of Syracuse is said [λέγεται] to have remarked about Archimedes, who had built a three-masted vessel which Hieron had ordered made for sending to King Ptolemy of Egypt. When all the Syracusans together were unable to launch it, and Archimedes made it possible for Hieron alone to move it down to the shore, he exclaimed, in his amazement: "From this day forth we must believe everything that Archimedes says." Tradition has it [φασιν] that Gelon made the same remark when, without destroying the crown that had been made, Archimedes discovered the weight of each of its component materials. (1.63, trans. Glen Morrow)

Although the four sources under consideration here span some five hundred years, range across several genres, and differ in detail and emphasis, every one of them invokes tradition at some point for all or part of the story.[25] Vitruvius tells the bath story without any reference to sources, but, as we have seen, he introduces Archimedes' method of solving the crown problem with "it is said" (*dicitur*). Plutarch presents the entire story as traditional (φασιν) and expects his readers to be familiar with the part about the measurement of the crown. Favinus introduces the story with "they say" (*ferunt*), and, as we have seen, Proclus invokes tradition when reporting the reactions of royalty to Archimedes' achievements. No one names a specific source for this anecdote, which could have been over two hundred years old when Vitruvius wrote his version.

This consistently traditional aspect of the story deserves more attention, because the story's repetition by the tradition is related to Archimedes' own repetitive speech. What Archimedes actually said, according to both Vitruvius and Plutarch, was not a simple εὕρηκα, but εὕρηκα, εὕρηκα, at least. Vitruvius says, "he called out again and again in Greek, 'I have found! I have found!'" (*identidem graece clamabat* εὕρηκα, εὕρηκα), and according to Plutarch, Archimedes went off saying εὕρηκα "over and over" (πολλάκις). To say εὕρηκα on finding something was not unusual (how else was a Greek to say "I have found"?); what was unusual was to say it repeatedly (and, of course, to do so while running through the public streets naked and dripping from a bath).[26] Vitruvius's markedly preserving and transmitting the Greek makes sense if εὕρηκα, εὕρηκα was already a famous exclamation. Otherwise he could just as well have stopped after the paraphrase *significabat clara voce invenisse, quod quaereret* (he indicated in a loud voice that he had discovered what he was seeking), or he could have rendered εὕρηκα into Latin: *inveni! inveni!* By Plutarch's time, εὕρηκα, εὕρηκα was an expression famous enough to be parodied (with βέβϱωκα, "I have eaten," and πεφίληκα, "I have kissed")—although Plutarch makes sure that his readers will understand his parody by telling Archimedes' story first. Moreover, this famous and repeatable piece of language became famous *because* it was repeated: Plutarch's Archimedes, who goes along saying εὕϱηκα "over and over," has said these words over and over before, in Vitruvius and also in the other (now lost) writers aligned behind the expression "as they say."[27]

Plutarch, indeed, has good reasons to invoke tradition, for he wants to draw as strong a contrast as possible between Archimedes, who both says εὕϱηκα, εὕϱηκα many times and says it in the many sources that make up the tradition, and the glutton and gallant, "no one [of whom] have we heard . . . shouting [οὐδενὸς δὲ ἀκηκόαμεν βοῶντος], though sensualists unnumbered have existed in the past and are with us now." Plutarch's assertion, that we have heard no glutton or gallant cry out, works on two levels. One reason no one has heard a person doing this is because the pleasure derived from food and love is less than that derived from mathematics. Therefore there is no initial cry—no single expression, even—of βέβϱωκα or πεφίληκα (to say nothing of βέβϱωκα, βέβϱωκα or πεφίληκα, πεφίληκα). A second reason no one has heard a person doing this is because such people accomplish nothing that justifies quoting their words. They neither shout nor earn the fame that would produce and preserve a story of their shouting. Plutarch must have been right, for βέβϱωκα and πεφίληκα have not, like εὕϱηκα, given names to towns, colleges, household appliances, and scholarly journals.[28] Thus, in Plutarch, the reference to tradition serves as both an editorial remark of the writer and a component of the story, and "they say" (φασιν) is thematically linked to Archimedes' famous utterance.

Of course, we do not know how Archimedes had his initial insight into specific gravity; if he really did spring from his bath and run home, naked, crying, εὕϱηκα, εὕϱηκα; or, if he did say that word, how many times he said it. We do not know Vitruvius's sources for the story (Varro and Cicero are possible suspects), or Plutarch's, or those of the other versions. But Vitruvius represents εὕϱηκα, εὕϱηκα as language that is already famous, already allusive. He points out that Archimedes said it, that he said it in Greek, and that he said it over and over. Archimedes' εὕϱηκα, εὕϱηκα became the English *eureka* because Vitruvius and Plutarch wrote that Archimedes reused it. What is generally lost in English usage is the initial—and, as we have seen, crucial—repetition.[29]

Let us return to the joyous, naked, and running Archimedes of Vitruvius, for there are other ideas associated with this striking image and, consequently, other ways in which Roman readers in the early Augustan Age might have re-acted to it. It might have brought to mind, for educated Romans as much as for educated Greeks, other examples of a traditionally Hellenic mental abstraction, such as the story in the *Theatetus* of Thales' tumble into a well or the account in the *Symposium* of Socrates' abrupt stopping to think.[30] Yet Archimedes does not fit the image of the Greek "intellectual" depicted in sculpture, who, when portrayed as thinking hard, is often shown seated with furrowed brow.[31] In addition, Archimedes' public nudity might have appeared memorably scandalous to Roman eyes, because the Roman sense of modesty extended even to exercising naked, a practice that appeared to Romans of the first century BCE as markedly Greek.[32]

Roman readers, then, might have interpreted Archimedes' nakedness and gait not only as signs of his Hellenic intellectual abstraction and unconcern about public nudity but also as indications that he lacked the ideal Roman quality of gravitas. Catherine Edwards and Anthony Corbeill have emphasized the importance of a person's dress and bearing, including his gait, in elite Roman society.[33] Corbeill points to the Roman "aversion to displaying emotional extremes publicly," for such display undermines gravitas.[34] Since observers can "read a person's internal moral makeup" by observing such externals as bodily control, even a walk that is too fast threatens one's hold on this elite virtue.[35] To readers situated in such a social climate, a joyous Archimedes running naked in public is both scandalous and laughable.[36] His body language, to say the least, is not that expected of the elite Roman.[37]

Indeed, Archimedes' movements can also be read within another, more specific interpretive context, that of the comic stage and its conventions. With the image of Archimedes leaping joyously from the bath, a socially "low" figure from the stage comes into view. In the *Second Philippic,* Cicero repeatedly shows how Marc Antony violates the social norms of the Roman elite. When he criticizes Antony's consumption of Pompey's estate, he casts him as a character in a mime: "so then, when he had unexpectedly drenched himself in that man's wealth, he leaped about with joy, a very character from a play, suddenly rags to riches."[38] Archimedes' leap from the tub (*exsiluit gaudio motus*), an open and uncontrolled display of emotions, thus dramatically demonstrates his low status, at least to an audience steeped in Roman social values.

Moreover, in his joyous running, Archimedes resembles yet another "low" stock character, that of the running slave, or *servus currens,* taking news to his master.[39] Whether the news he brings is good or bad, the *servus currens* displays his emotions openly as he runs along delivering his lines.[40] This is not to say that the "Eureka" story originated on the comic stage or ever appeared there (although speculation along these lines is tempting), but Archimedes' procession through the city could have been read by first-century Romans as a comic version of the victory lap, one casting Archimedes as slave and Hieron as master.

The element of comedy in this image raises the issue of social class, even as it makes the antithesis between abstracted intellectual and self-absorbed athlete all the more pointed. Reviel Netz has pointed out how difficult it is to pinpoint the social status of ancient mathematicians, yet he has also been able to argue convincingly that Archimedes came from a family of artisans.[41] Cicero called Archimedes "lowly" (*humilis*), and Silius Italicus refers to him as "bare of resources" (*nudus opum*); but in Plutarch, Archimedes has servants, addresses Hieron without deference, and is said to be related to the royal family.[42] In addition, as Netz points out, Archimedes was literate, had leisure and access to writing materials, traveled, and carried on a correspondence with the intellectual elite throughout the Mediterranean world. The issue I wish to address, then, is not what Archimedes' class was, since it seems to vary with the writer, but how presenting him as of a particular class serves the interests of the particular writer who portrays him.

Archimedes' class becomes especially meaningful if we approach Vitruvius's preface as a piece of dedicatory literature, an address from architect to Augustus.[43] When we do so, we note, first, that Vitruvius introduces the anecdote as Hieron's story: Hieron was "increased by regal power" (*regia potestate auctus*); Hieron wanted to dedicate the crown; Hieron saw to the letting out of gold, was outraged at being cheated, and asked Archimedes to solve the problem. All the attention focused on Hieron—none of which appears in Plutarch's version—makes Archimedes a paradigm of genius at the command of the ruler. (It also hints at a Hellenistic source for the written version of the story.) The "Eureka" story, then, stands out from the other examples in Vitruvius's preface as the only one portraying a relationship analogous to that of Vitruvius and Augustus. Yet Vitruvius makes it clear that Archimedes and Hieron represent extremes in a way in which he and Augustus do not. He establishes a set of antitheses between Hieron and Archimedes: king versus subject, regal offense versus intellectual joy, the king's ignorance (*neque inveniens qua ratione id furtum reprehenderet*) versus the scientist's knowledge (*cum eius rei rationem explicationis ostendisset*), the king "increased with regal power" (*regia potestate auctus*) versus Archimedes portrayed as a comic slave. Like Hieron, Augustus is *auctus*, but unlike Hieron, he is not "increased with kingly power." In fact, he would later pointedly claim to govern with *auctori-*

tas rather than *potestas*.[44] Like Archimedes, Vitruvius serves his patron, but he does so indirectly, through the intermediary of the *De architectura*, which gives him authority that he would lack in person.[45] Writing makes it possible for Vitruvius, who elsewhere presents himself as a man of humble background and unimposing stature, to enter into a relationship with Augustus but keep his dignity.[46]

Vitruvius's joyous Archimedes presents yet another challenge to aristocratic values. Vitruvius presents the "Eureka" story as the paradigmatic account of *inventio*, which is itself the concept central to the preface of book 9.[47] Archimedes is clearly Vitruvius's most important exemplum, for he tells this story at greater length and in more detail than any of the others. It is, moreover, the one exemplum in this preface that reflects the chronology of discovery: first the initial insight, then the systematic working out of the problem.[48] It stands out from the others for the attention given to Archimedes' emotional response.[49] Finally, it alone of Vitruvius's examples quotes its subject directly—and in his own language. The result is that Archimedes is always present in this story; he always cries, εὕρηκα, εὕρηκα, any time that anyone anywhere reads Vitruvius. The anecdote's rhetorical effect, then, is to re-create vividly for all readers, always, the moment of discovery.

As we have seen, Archimedes' joyous impulse was to give vent to the fact of his discovery in a manner that showed utter disregard for his body. Yet that body, to which the reader's gaze is irresistibly drawn as a body (*corpus*), has meaning that extends beyond this story to embrace Vitruvius's concept of his entire project.[50] As Vitruvius explains at length in book 3, the proportional measurement that gives symmetry to architecture begins with the proportional measurement of man—his height, the length of his head, his foot, his arm, his hand.[51] If we reread the "Eureka" story in this context of man as measure, Archimedes in his bath represents the discovery of yet another way in which to use the body as a means of measuring. Archimedes noticed that "however much of his own body settled into the tub, precisely that much water flowed out of it."[52] His body, then, is neither something to improve, like the athlete's, nor a social signifier to be imbued with or robbed of dignity, like the elite Roman's, but simply a mass that displaces a certain volume of water and thus serves as part of the measuring device by which he makes his discovery.

With his discovery, Archimedes has found a way to vouch for the truth of a statement, not the claim of the contractor who cheated Hieron, but the evidence (*indicium*) of the person who alleged the fraud.[53] Moreover, his method of measuring is infinitely applicable: it can detect the genuine as well as the false—hence Archimedes' joy even at detecting a theft that insulted gods and king. Thus the comic Archimedes, rejoicing at his discovery, presents readers with a new perspective on value. He leaps with joy, like Cicero's mime figure in *Philippics* 2.65 who has gone "suddenly rags to riches" (*modo egens repente dives*); yet Archimedes has not discovered riches. Indeed, he has discovered the opposite: the absence of gold, which is the traditional standard for wealth. This discovery is cause for joy only in the eyes of a person who places a higher value on knowledge than on wealth. Seen from this perspective, our low-born character appears to have discovered something more valuable than the stolen gold, something even King Hieron himself could not discover, a *ratio*—not just a "method," but one of the ruling principles that tie the universe together.

Leslie Kurke has pointed out the "repeated and emphatic mention in aristocratic poetic texts of the sixth and fifth centuries B.C.E. of the weighing, refining, and testing of precious metals."[54] It is a topos in didactic poetry (e.g., Theognis 77–78, 1104 a–6), one adapted in epinician poetry (e.g., Pindar *Pythian* 10). Kurke observes, "What all these passages have in common is that they come from aristocratic texts and specifically from contexts that mark an in-group of *hetaireia* or *xenia*."[55] Gold is a symbol of aristocratic community and aristocratic exchange. Kurke has also pointed out the symbolic link between gold and sovereignty.[56] Although Hellenistic Syracuse and Augustan Rome are far from the world of archaic Greece, Vitruvius invokes this world by referring to the rewards given athletes at the games; moreover, Archimedes' status and relationship to Hieron, the nature of Hieron's rule, Vitruvius's anxiety about status and pay, and the fact that he is addressing his ruler, Augustus, all suggest that it is worthwhile to map Archimedes' story onto the conceptual system laid out by Kurke. When we do so, we can see the contractor's substitution of silver for gold not simply as an act of impiety toward a god but as a test of Hieron's sovereignty as well. Hieron, beloved ruler and loyal ally of Rome, was technically a tyrant in that he seized power by extralegal means. The phrase *regia potestate auctus* at the beginning of the anecdote places the crown story not long after his coming to power, when his grasp on it might not yet have been secure. As gold tested against the touchstone has real value, so does the tried and true comrade in civil strife.[57] Archimedes' discovery of a means for determining the purity of metal replaces the touchstone of aristocratic discourse. His genuinely spontaneous behavior makes him an ideal "trusty comrade" (*pistis hetairos*), one who has no "trickery within."[58] This story, then, recasts the discourse of archaic aristocracy so as to show that men whom Vitruvius presents as nonaristocrats, such humble men as Archimedes and himself, can serve their sovereigns as trusty com-

rades. By imposing Kurke's "language of metals" onto the relationship between Archimedes and Hieron and projecting it onto Vitruvius and Augustus, we can read the "Eureka" story as an illustration of the advantages that
accrue to a ruler from having as a loyal associate a member of the intellectual
elite, however lowly his birth.

My intention in this chapter is not to enter deeply into issues of Augustan patronage—some of my observations simply complement arguments made by
others, especially Indra McEwen and Mark Masterson—but, rather, to make
a point about how familiarity obscures the obvious (Archimedes does not
even appear in McEwen's book). As a modern expression, *eureka* is so familiar as to be taken for granted. Returned to its context, it reveals an image of
Archimedes as a figure of the intellectual athlete and a topos for discovery.
Revealing this topos, in turn, informs us more about the rhetorical effects of
his story on the texts in which it appears. We can see that Plutarch presents
the "Eureka" story so as to support doubly his diatribe against Epicureanism.
Likewise, we can see that Vitruvius has presented his version so as to make it
pertinent particularly to his argument about the rewards due writers and
generally to the greater social context of the preface, that of citizen-writer addressing the *princeps*. As I turn in this book to other parts of the anecdotal
tradition, I will return to some of the ideas raised by this story. The element
of ridicule associated with Archimedes, his ability to call forth laughter,
resurfaces and contributes in different ways to the representation of his genius as un-Roman. References to tradition will reappear and guide the interpretation of other anecdotes. Other artifacts associated with Archimedes will,
like the crown, be associated with radical shifts in assessing value. Finally, the
"Eureka" story is important for what follows because it provides a paradigmatic image: it places Archimedes, his body and his voice, at the center of a
"primal scene" of discovery, one that sets physical limitations against mental
freedom, permanence against impermanence, and the world of the senses
against the world of ideas.

CHAPTER 2

CICERO AT ARCHIMEDES' TOMB

A PAINTING BY BENJAMIN WEST, in the Yale University Art Gallery, shows a group of figures in classical dress in a Mediterranean setting. Smoke rises from a volcano in the background. Almost all the adult men are bearded, by which West conveys that they are Greek; one is rather portly. On the right, a slender, clean-shaven man, conspicuous in a gleaming toga, gestures energetically with one hand while pointing at a ruined monument with the other. In the foreground, at the lower right, two half-clothed men, their limbs arranged in lively diagonals, wrest branches away from the ruin, while a shadowed figure behind it brandishes a sickle. All the movement in the painting is directed toward this monument; but it is the bright figure of the gesturing man that catches and holds the eye.[1]

This figure represents Cicero, for the painting takes its subject matter from the story Cicero tells, in the last book of the *Tusculan Disputations,* about his discovery of Archimedes' neglected tomb. While the details of Archimedes' bath have been obscured by the familiarity of the cry "Eureka!," the story of his tomb has, as it were, fallen out of the sight of scholars, into a gap between two sets of biographies. Unlike West and a few other painters of the late eighteenth and early nineteenth century, classical scholars generally ignore this anecdote, except for those who, taking it at face value, refer to it when they discuss Archimedes' death or Cicero's career.[2] Yet it merits attention on its own account for a number of reasons: first, because of what it has to say about memory and commemoration; second, because it brings together several of the dialogue's major themes and, in doing so, acquires a meaning greater than their sum; and third, because of the relationship it establishes between two men who appear at first glance to have had very little in common, the elderly Greek mathematician and the youthful Roman advocate.

What follows explicates this artifact in several interrelated ways. It examines briefly the structure of the anecdote and the way it fits into its various frameworks—autobiographical, discursive, and intellectual. Then, by looking at the anecdote's rhetorical features, it shows how Cicero constructs a new textual monument from the fragments of a physical one. After that, delving more deeply into matters of context, it considers this written monument's symbolic function in the *Tusculan Disputations* as a whole and its role in two related arguments presented by the dialogue: that the soul is immortal and that virtue is sufficient for living happily. Parts of this last section bolster points made in other discussions of exempla in Roman philosophical debate, but by approaching these reminders of past virtues and vices in their most concrete form—that of monuments—I hope to illustrate further the role played by memory and authority in this particular discussion of moral issues.[3] Finally, I want to examine the anecdote in the larger context of Hellenistic Greek and Roman cultural interaction. According to Cicero, the *Tusculan Disputations* were rhetorical exercises on philosophical topics, and Paul MacKendrick has observed that the use of rhetoric for philosophical ends is "the heart of Cicero's originality."[4] Here, at least, Cicero's originality stems from the features of his writing that might be called the most Roman: the personal, the concrete, and the exemplary.

The fifth book of the *Tusculan Disputations* argues that virtue is sufficient for living a happy life (*virtutem ad beate vivendum se ipsa esse contentam*, 5.1), a climactic argument whose foundations were established in the first four books.[5] Here, after arguing his point by syllogism and recalling discussions that showed the wise man to be happy, Cicero invokes three pairs of contrasting exempla.[6] He begins with the Romans Laelius and Cinna: Laelius's moral standards were impeccable, although he lost a bid for the consulship; Cinna won four consulships but murdered noble men and was killed by his own soldiers. There follow Catulus, who died at Marius's orders, and Marius, who killed him. Last of all, Cicero introduces a non-Roman exemplum, one that receives its counterpart only after extended elaboration: Dionysius the Elder, tyrant of Syracuse.[7] This is the Dionysius who suspended the famous sword over the neck of Damocles to convey a sense of the constant fear in which he lived. After telling the story of Dionysius, Damocles, and the sword in some detail, Cicero asks, "Does not Dionyius seem to have made it sufficiently clear [*satisne videtur declarasse Dionysius*] that there can be nothing happy [*beatum*] for the person over whom some fear always looms?" (5.62). After a brief account of Dionysius's admiration of Phintias and Damon, whose friendship transcended the fear of death, Cicero returns to the tyrant's wretchedness and then tells the story about Archimedes' tomb.

(A) Non ego iam cum huius vita, qua taetrius miserius detestabilius excogitare nihil possum, Platonis aut Archytae vitam comparabo, doctorum hominum et plane sapientium: ex eadem urbe humilem homunculum a pulvere et radio excitabo, qui multis annis post fuit, Archimedem. (B) cuius ego quaestor ignoratum ab Syracusanis, cum esse omnino negarent, saeptum undique et vestitum vepribus et dumetis indagavi sepulcrum. (C) tenebam enim quosdam senariolos, quos in eius monumento esse inscriptos acceperam, qui declarabant in summo sepulcro sphaeram esse positam cum cylindro. (D) ego autem cum omnia conlustrarem oculis—est enim ad portas Agragentinas magna frequentia sepulcrorum—animum adverti columellam non multum e dumis eminentem, in qua inerat sphaerae figura et cylindri. (E) atque ego statim Syracusanis—erant autem principes mecum— dixi me illud ipsum arbitrari esse, quod quaererem. (d) inmissi cum falcibus multi purgarunt et aperuerunt locum. quo cum patefactus esset aditus, ad adversam basim accessimus. (c) apparebat epigramma exesis posterioribus partibus versiculorum dimidiatis fere. (b) ita nobilissima Graeciae civitas, quondam vero etiam doctissima, sui civis unius acutissimi monumentum ignorasset, nisi ab homine Arpinate didicisset. sed redeat, unde aberravit oratio: (a) quis est omnium, qui modo cum Musis, id est cum humanitate et cum doctrina, habeat aliquod commercium, qui se non hunc mathematicum malit quam illum tyrannum? si vitae modum actionemque quaerimus, alterius mens rationibus agitandis exquirendisque alebatur cum oblectatione sollertiae, qui est unus suavissimus pastus animorum, alterius in caede et iniuriis cum et diurno et nocturno metu.[8] (5.64–65)

[Now with the life of this man, than which I can imagine nothing more vile, wretched, and loathsome, I shall not compare the lives of Plato or Archytas, men learned and clearly sages: instead I shall summon from his dust and measuring rod a lowly, insignificant man from the same city, who lived many years later, Archimedes. When I was quaestor I sought out his grave, which was then unknown to the Syracusans (as

they utterly denied its existence), and found it surrounded on all sides and overgrown with thornbushes and thickets. For I remembered certain lines, inscribed, as I had heard, upon his memorial, which stated that a sphere together with a cylinder had been set up on the top of his grave. Accordingly, after surveying all around (for there is a great throng of graves at the Agrigentine Gate), I noticed a small column standing out a little above the bushes, on which there was the figure of a sphere and a cylinder. And so I said immediately to the Syracusans—their leading men were with me—that I thought it was the very thing that I was seeking. A number of men sent in with sickles cleared the place, and when an approach had been opened, we drew near the pedestal facing us; the epigram was visible, with the latter parts of the verses half eaten away. Thus the most famous city of the Greek world, once even the most learned, would have been ignorant of the memorial of its most keen-witted citizen, had it not learned of it from a man of Arpinum. But let my discourse return to the point from which it has strayed. Who is there who has anything to do with the Muses, that is with humanity and learning, who would not prefer to be this mathematician rather than that tyrant? If we look into their manner of life and employment, the mind of the one was nourished by seeking out and pondering theories, accompanied by the delight he gave his own cleverness, which is the sweetest sustenance of souls, that of the other in murder and wrongdoing, accompanied with fear both day and night.]

The Structure and Context of the Anecdote

The *Tusculan Disputations* was one of several works of moral philosophy that occupied Cicero in the years 45–44 BCE. Cicero explains his reasons for writing the dialogue in its opening sentences, saying that he will try to translate Greek philosophy into Latin, now that he has been released from his duties as advocate and senator (1.1).[9] According to Cicero, the project is necessary because although Greece no longer surpasses Rome in customs, morals, or government (to say nothing of war and oratory), it still surpasses Rome in theoretical knowledge (*doctrina*, 1.3).[10] Cicero, then, explicitly locates the dialogue within the broad intellectual and historical framework of the ongoing Roman endeavor to appropriate Greek culture and, at the same time, within the narrower autobiographical framework of his political and intellectual career. Indeed, the opening remarks place the dialogue on the threshold that marks a point in the continuum extending from a Roman's public service to his private life, and they commemorate the act of passing over that threshold.[11] I will return to this context later.

The passage itself is a rhetorical figure, a *digressio,* set off from the discussion of the happy life by an introduction and return (*Non . . . comparabo. . . . Sed redeat, unde aberravit oratio*).[12] The impression that the story returns at the end to the very place where it left the main argument is reinforced by its ring composition.[13] The digression departs from the original topic—the happy versus the wretched life, or *vita* (A)—and moves on to Archimedes' tomb, with an implied contrast between Cicero's curiosity and knowledge and the Syracusans' ignorance and lack of curiosity: when quaestor in Sicily (*ego quaestor*), Cicero searched out (*indagavi*) Archimedes' tomb, which he says was unknown until then (*ignoratum*) by the Syracusans (B). The digression then moves on to the verses retained in Cicero's memory (C), the discovery of the monument itself (D), Cicero's proclamation of his discovery (E), and the approach to the monument (d). After returning to the verses (c), the digression ends with Cicero pointing out that the most noble and once most learned of Greek cities would have been ignorant (*ignorasset;* cf. *ignoratum* in [B]) of the monument of its most ingenious citizen, had it not learned of it from him, a man of Arpinum (b). He then returns to the main discussion of the happy versus wretched life (*Si vitae modum . . . quaerimus*) (a).

The digression, of course, is not necessary to the immediate argument. Indeed, its elimination leaves no gap, since Cicero compares the lives of Dionyius and Archimedes only at (a), after the digression about the tomb.[14] The digression removed, the main argument flows smoothly from the introduction of Archimedes (A) to the happiness of his life of scientific inquiry compared with the misery of Dionysius's paranoia (a). We have, then, an anecdote distinct from the main argument, presented as an eyewitness account of an incident by one of the speakers in the dialogue (*ego quaestor . . .*), and centered on the moment of discovery. Yet the digression is embedded in a climactic pair of contrasting exempla, and this pair appears in the middle of the climactic argument of a theoretical discussion of ethics ostensibly in the Greek manner (*Graecorum more,* 1.7).[15]

Cicero did not need to tell this story about the tomb in order to demonstrate that Archimedes was happier than Dionysius. Indeed, it hardly serves that purpose, at least not explicitly, nor does it evoke the reader's assent in the same way as do the references to Laelius's good character and Catulus's better death. Cicero might have inserted an anecdote illustrating Archimedes' happy lifelong preoccupation with mathematics, the very idea that he takes as given in his rhetorical question "Who is there who has anything to do with the Muses . . . ?" Or Cicero might have shown a vivid image of Archimedes happy at work and oblivious to his own impending death, as he does at *De finibus* 5.50. In fact, Cicero's declaration that he will rouse Archimedes from his drawing board and pencil might well make a Roman audience recall the other famous instance of a Roman's disturbing the mathematician at his work: during the sack of Syracuse, in the middle of the Second Punic War, Archimedes, as the story went, was so intent on studying his geometrical figures that he ordered a Roman soldier to leave them alone and was killed for it. Indeed, until Cicero deflects the course of the digression to his discovery of the tomb, the contrast with Dionysius appears to be leading precisely to this point. Fear of death made Dionysius wretched, and by suspending the sword over Damocles' head, he communicated clearly (*declarare*) how it was impossible for a tyrant to enjoy pleasure. The antithesis is neat and clear: the tyrant Dionysius, fearing for his life, and the humble private citizen Archimedes, oblivious of the danger, are united by the symbol of impending death, the metaphorical sword hanging over every hated tyrant's head and the very real weapon suspended over that of the unconcerned Archimedes.[16]

Archimedes' death may have been a sore point for the Romans, so there is that possible negative reason for diverting the discussion from it.[17] But we need not speculate too much here about what Cicero did not say. The internal logic of the *Tusculan Disputations* presents compelling reasons why the speaker in the dialogue should place Archimedes and his death in the background and bring his own image into the fore.

The Rhetoric of the Anecdote

Even the most casual reader of the *Tusculan Disputations* will observe that Archimedes is not the central figure in this story. After Cicero says that he will compare Dionysius's life to that of Archimedes, he refers to Archimedes only by pronouns in the genitive, as the unnamed possessor of a tomb (*Cuius . . . sepulchrum . . . in eius monumento*). Instead, the rhetoric of the passage calls attention to Cicero and his search: Cicero is the subject of the first four main verbs, all of them active (*indagavi, tenebam, animum adverti, dixi*); he is also subject of three subordinate ones (*acceperam, conlustrarem, quaererem*). The repetition of the emphatic personal pronoun *ego* draws more attention to the aggressively seeking Cicero (*ego quaestor . . . indagavi sepulcrum. . . . Ego autem . . . animum adverti columellam. . . . ego statim Syracusanis . . . dixi*). In fact, Cicero alone is the subject of verbs denoting inquiry; after he announces his findings (*Syracusanis . . . dixi*), the Syracusans join him as the subject of the verb *accessimus*. Other details reinforce the sense of Cicero's lively and aggressive curiosity: the title *quaestor* offers a weak pun, since it is derived from the verb *quaerere*, "to search," which describes both Archimedes' life of inquiry (*alterius mens rationibus agitandis exquirendisque alebatur . . .*) and Cicero's (*quaererem, quaerimus*). In addition, the verb *indagare*, "to track," introduces a hunting metaphor used elsewhere of philosophical inquiry.[18]

The monument, too, is active, if less so. The verses on the column are the subject of verbs denoting communication and visibility (*declarabant* and *apparebat*). This dual emphasis on Cicero searching and the verses signaling suggests that the purpose of the anecdote is less to compare Archimedes and Dionysius than to bring together Cicero and Archimedes' tomb and thus, indirectly, Cicero and Archimedes, a juxtaposition captured nicely in the word order of the first sentence: *Cuius ego quaestor . . . indagavi sepulchrum.*[19] This juxtaposition picks up a motif from earlier in the *Tusculan Disputations*, where Cicero writes that Themistocles, when asked why he spent his nights wandering about the city, replied that the trophies of Miltiades kept him awake. By juxtaposing one great man and the reminder of another, the story presented a vivid and memorable image of rivalry between the living and the dead.[20] In placing himself at Archimedes' tomb, Cicero himself joins Themistocles in confronting a reminder of the dead, and the nature of that confrontation suggests that this story, too, represents some kind of comparison and perhaps rivalry. In the case of Cicero and Archimedes, however, the juxtaposition of man and monument becomes more complex and intriguing when we realize that the anecdote that juxtaposes Cicero and Archimedes also replaces Archimedes with Cicero as the contrast to a miserable tyrant. What is this digression doing? For answers, let us start by turning to the monument itself, as it appears in Cicero's text.

Cicero's passage is the only description of Archimedes' tomb by someone who claims to have seen it. It is also the most detailed description, although Plutarch writes in his life of Marcellus that Archimedes, wanting to commemorate his greatest achievement, asked that a cylinder and a sphere, together with the formula for the ratio of their volumes, be put on his tomb.[21] Plutarch and the *Tusculan Disputations* agree on the basic points: there was a tomb and, above it, a figure of sphere and cylinder, together with an inscription. The *Tusculan Disputations* describes a column, with the cylinder and sphere on top, resting on an inscribed base.[22] We do not know precisely how the sphere and cylinder appeared on Archimedes' monument: D. L. Simms argues that it was a three-dimensional stone figure, with part cut in the form of a cylinder and the rest cut away to show the inscribed sphere; Reviel Netz, pointing to other illustrations of Archimedes' thought, argues that it was an inscribed image—a diagram.[23]

The very fact that there is debate about its appearance ought to remind us that although Cicero gives a fairly detailed sketch of the discovery of the tomb and himself as eyewitness, it is part of a digression; this digression, moreover, is a rhetorical description offered in the voice of the Cicero who is a participant in this dialogue. This first-person speaker remembers his younger self looking for the tomb; and that younger, remembered Cicero in turn remembers verses that he heard earlier (*Tenebam. . . . acceperam*). These several Ciceros, all speaking in the first person, are easily telescoped into one subject. Yet time separates them, tense marks their separation, and memory joins them together. I shall return to this point later but for now note simply that each Cicero depends for his existence on the recollection of the later one.

Moreover, unlike the Syracusan dignitaries, the audience of the *Tusculan Disputations* does not see the monument directly; it sees what the narrating Cicero chooses to show and does not see the features that he chooses to omit. The digression neither describes how the cylinder and sphere were constructed nor quotes the verses on the column base. Cicero instead refers vaguely to the formula that interprets the figure, which also is described in general terms. Thus the digression, while giving Cicero's audience a view of the discovery of the monument that was lost and found again, permits only limited access to the monument itself. This restriction is deliberate and important. Cicero alone stands behind—or perhaps we should say in front of—this monument: his memory led him to find it; his words preserve it; his *auctoritas* as an autobiographical first-person narrator guarantees it to be what he says it is.

As Cicero represents the grave marker, it is failing. According to Cicero, the verses at the base of the monument referred to the presence of a cylinder and sphere atop the *columella* above.[24] The monument thus described is a self-sufficient system, with verse interpreting image and image illustrating verse. The grave marker draws attention to the ratio of cylinder to sphere and, in doing so, draws attention away from the man whom it was built to commemorate. Yet even Archimedes, that inventor of gadgets, could not produce a completely self-sufficient monument, for when Cicero finds it, the system is beginning to fail: time has eaten away the last half of each of the lines of verse on the base; brush has grown up to obscure the *columella* above. As the monument deteriorates, it loses its ability to commemorate Archimedes and his achievement, for if the verses are worn away, no one will be able to read the epigram, and if the brush grows high enough, no one will be able to see the cylinder and sphere. Cicero's description, moreover, places emphasis on the faintness of the signal emitted by the monument. The verses Cicero once heard are described with diminutives (*senariolos, versiculorum*); the column, when he finds it, is a small one that barely peeks out above the surrounding brush (*columellam non multum e dumis eminentem*).[25]

The monument's deterioration, however, allows Cicero to become a part of the system commemorating Archimedes and, consequently, to forge a set of links between his own life story and that of Archimedes, which point, as we shall see, to particular moral conclusions. As Cicero says, he found the tomb because he remembered the verses he had heard (*Tenebam . . . acceperam*). He brought to his search prior knowledge of the monument, which had been gained through poetry and stored in memory. While the monument itself originally represented the reciprocity of figure illustrating formula and formula interpreting figure, its deterioration allows Cicero to become part of another reciprocal process. The verses, which had been originally seen on the monument and repeated at some point to the younger Cicero, remain in his memory; his memory of those verses leads him to rediscover the monument from which those verses came. Moreover, the process beginning in the material world with the verses on the monument begins in the digression with Cicero's actions (*Tenebam . . . acceperam*); that is, the remembered verses help Cicero find what the remains of the inscribed verses confirm that he has found. This movement of knowledge from the tomb to Cicero's memory and back to the tomb includes Cicero, who is necessary for the process, and the Syracusans, with whom he shares his findings, but excludes Archimedes. As a first-person narrator, then, Cicero offers a restricted, or focalized, view of himself and of the monument, with no glimpse of Archimedes after the early reference to the *homunculus* engrossed in figures in the dust. Even at the moment of discovery, Cicero, as he reports it, said, "I thought it was the very thing that I was seeking," rather than, "I thought it was Archimedes' tomb." For a story introduced as one about the inquisitive (and, by implication, happy) life of Archimedes, it is a fine story about the inquisitive (and, by implication, happy?) life of Cicero, who has usurped Archimedes' role as discoverer (*ego quaestor . . . indagavi*).[26]

The Anecdote's Symbolic and Rhetorical Function

At the simplest level of interpretation, we can say that the anecdote about Archimedes' monument commemorates Cicero's discovery of it. Yet in doing so, it intertwines the lives of two men, both known for practical and theoret-ical accomplishments, the Roman statesman and philosopher and the Greek inventor and mathematician. Let us bring the autobiographical framework back into play: Cicero discovered the monument in 75 BCE, when he was holding the quaestorship, the first rung in the traditional Roman political career. He was married to Terentia and the father of a young daughter. By the time he wrote the *Tusculan Disputations* in 45 BCE, one man, Julius Caesar, was at the head of the state; Caesar had pardoned Cicero for siding with Pompey in the civil war; and Cicero had withdrawn from active political life. Cicero's marriage with Terentia had ended, and, to make matters worse, his daughter Tullia had died at his estate in Tusculum, this dialogue's setting and a place to which Cicero could not bring himself to return for some time after her death.[27] Discouraged about his role in Roman political affairs and griev-ing for his daughter, Cicero turned to philosophy and produced most of his philosophical works in this year and the next.[28] Thus, in recollecting his dis-covery of the monument, Cicero looks back to the beginning of his political career from the vantage point of thirty years later, a point that he perceives to be its end.[29] The recollected discovery of Archimedes' tomb, then, links Cic-ero's early political life to his present and allows a return to the past, a past now commemorated and reinterpreted by the anecdote as part of his life of inquiry as well as his political life.

The anecdote also contrasts the constant flux of the physical world with the power of human memory, for Cicero's representation of the monument questions its endurance over time. In the years between Archimedes' death and Cicero's quaestorship, the last half of each line has been eaten away.[30] Thus, in Cicero's reconstruction, the tomb becomes a site where the in-evitable decay of monuments meets the tenacity of human memory, and both are set against the eternal truth discovered by the mathematician, the ratio of the volumes of cylinder and sphere.[31]

The story of Archimedes' tomb, then, brings together several mnemonic processes: the digression moving out of the main argument and returning to the place it began; the monument itself as the text represents it, a self-refer-ential reminder, with figure and formula explicating each other; Cicero's rec-ollection of the beginning of his political career at what appears to be its end, as well as his incorporation of that beginning into his representation of a life devoted to inquiry; the movement of the verses, which came from the mon-ument and lead back to it via Cicero's illuminating memory; the return of *doctrina,* which had departed from the Greeks and now returns to a Syracuse that was once the most learned of cities (*quondam . . . doctissima*) but is now ignorant of even the reminder of its sharpest citizen, a return that it owes to the tenacious memory of a visiting Roman from a small Italian town (*nisi ab homine Arpinate didicisset*).[32] Behind all these acts of memory associated with Cicero's public life lie his return to philosophy as a consolation and his attempt to rewrite the memory of Tusculum by returning to it as the setting for the dialogue.[33]

For the Syracusan dignitaries accompanying Cicero, a clear path leads to the monument, and the monument points to a lost and greater past, back to Archimedes, the sack of Syracuse, and the time before the city fell into Roman hands. For the audience reading or listening to the *Tusculan Disputations,* the path leads to the digression, which points to Cicero and his discovery. This bring us to two more observations. First, this written *monumentum* brings together the beginning and end of the *Tusculan Disputations* by joining the arguments that the soul is divine and immortal and that virtue suffices for the happy life. Second, this neat rhetorical figure, this written monument, works both to help Cicero's audience remember a group of related ideas and to persuade it to act in a particular way.

Memory comes into play early in book 1, when Cicero uses monuments as material evidence supporting his argument for the immortality of the soul. The soul must be permanent, Cicero says, because everyone is concerned about what will happen after death. Why else, he asks, would people plant trees whose fruit they would not live to eat; why else would great men sow the seed of laws and public policy? "The very burial monuments [*ipsa sepulcrorum monumenta*], and the epitaphs [*elogia*]—what meaning have they except that we are thinking of the future as well as the present?" (1.31). The reference to Archimedes' monument and Cicero's memory would call to mind these earlier references to monuments and memory and their role in lending the *auctoritas* of the past to Cicero's argument.[34]

When Cicero argues for the divinity of the soul, he particularly invokes memory and invention to support his point (1.57–61). Cicero first discusses Platonic memory (anamnesis, the soul's recollection of its prior knowledge of the forms). Then he says that he finds regular memory even more wonderful—not the artificial technique of the *ars memoriae,* but regular human memory, the kind found in people who have mastered some higher branch of study and art.[35] In fact, precisely this kind of memory—that of the trained orator who recalls arcane literature (in this case, the lines of poetry) and uses it to prove a point—leads Cicero to Archimedes' tomb and is then set against the fallibility of a decaying monument.[36] For Cicero, as he narrates the story of the tomb, the memory of his own previous lives as student, orator, and statesman replaces the Platonic soul's memory of a previous existence (*recordatio vitae superioris*, 1.57). These are obvious points of connection, and they may at first seem to be the only ones, but when we look at how the written monument works rhetorically, we see that there are more and that they are more complicated.

First, when a writer pairs historical exempla, he presents his audience with a moral choice that is essentially a matter of sympathy: whose side does it prefer?[37] This is, of course, not a true choice, since the writer's sympathy clearly falls on one or the other side. Cicero could expect an audience of aristocratic and educated Romans to be aware of its own potential future as exempla to imitate or avoid. The audience members would feel sympathy and a sense of emulation in regard to the exemplum associated with a good "obituary." In a pair of contrasting exempla, monuments associated with the good example help tilt the argument in its favor by giving the person commemorated the authority of having done something worthy of sustained memory.[38] This fits nicely a very Roman way of thinking: men who contribute to the state or to humankind through their *virtus* are commemorated. It is natural to turn this around and infer that the monuments one sees commemorate men sufficiently endowed with *virtus* and concerned about posterity to contribute to society during their lives. Archimedes has left a monument worth finding, thus a memory worth preserving: therefore Archimedes was a citizen endowed with *virtus*. Dionysius, in contrast, has left only a city that has outlived its former glory, anecdotes illustrating his misery, and some bad tragedies.[39] Virtue, in fact, may not be self-sufficient, at least not for the happy afterlife; memory is also necessary. It is a compromise: remembered virtue resulting from a permanent contribution to the state or to human knowledge suffices for happiness.[40]

When Cicero redirects attention from Archimedes to himself, he takes Archimedes' place as the good exemplum in contrast to the tyrant Dionysius. Like Archimedes, Cicero is a man of practical and theoretical accomplishment, who contributes to his state through the force of his intellect.[41] It is possible that Cicero sees his defense of the Roman Republic by means of all the political contrivances he could muster as analogous to Archimedes' efforts at defending Syracuse with all the mechanical contrivances he could devise, and Cicero may see his withdrawal into philosophy (1.1) as analogous to Archimedes' concentration on his diagrams when all was lost. If Cicero replaces Archimedes, who replaces Dionysius? By this time, Cicero considered Caesar a tyrant.[42] Writing to Atticus, he had already compared his own use of his wit in an environment controlled by Caesar to Archimedes' ingenuity. One letter complained that he had to deliver a funeral oration for Cato in Caesar's presence, and the other said that he had solved the problem. In each case, he calls the matter a πρόβλημα Ἀρχιμήδιον (problem for Archimedes).[43] Whether or not Cicero intended his audience to see Caesar in the figure of Dionysius, hindsight shows us a clear picture of more swords, not just those hanging over the heads of Damocles and Archimedes, but those hanging over the heads of Caesar and of Cicero himself. The question, then, is not only which life is happier, Cicero's or Caesar's (and here Cicero, who has been trying to convince himself all through book 5 that virtue suffices for happiness, is particularly defiant), but also which obituary does one want?[44] The answer seems to be that of the defender of his state who, when all is lost, turns his thoughts to his figures in the sand.

Second, what kind of legacy ensures a conscientious heir? Archimedes was a private citizen and, Cicero claims, obscure (*humilem*); the physical traces of his memory were slight (*senariolos; versiculorum; columellam non multum e dumis eminentem*); yet Cicero sought and found his tomb, ensuring the continuity of his memory—but not by Archimedes' fellow citizens. It is the Roman Cicero who ensured, as young quaestor, and continues to ensure, as a mature ex-consul, commemoration for the dead Greek mathematician.[45] In remembering Archimedes, he pronounces him the most acute (*acutissimi*) citizen of the noblest and once most learned (*doctissima*) city, as if he were delivering a funeral oration, a kind of speech that "tends to celebrate accomplishments in terms of superiority over other members of society."[46] But it is not just the dead who tend to be described in superlatives. The same goes for heirs.[47] The Syracusans' ignorance about the location of Archimedes' tomb and their active denial of its very existence (*cum esse omnino negarent*) cast them as unworthy stewards of their inheritance. It was, after all, the heir's responsibility to see to the tomb's upkeep. In commemorating the dead, in taking care of the tomb, Cicero takes on the role of the dead man's true heir, the person who, as Cicero says elsewhere, comes closest to replacing the one who has died.[48]

Finally, Cicero rouses Archimedes just as he conjures up philosophy (cf. *excito* at 1.5 and 5.64): he brings light (*inlustranda; conlustrarem*) to both.[49]

> Philosophia iacuit usque ad hanc aetatem nec ullum habuit lumen litterarum Latinarum; quae inlustranda et excitanda nobis est, ut si occupati profuimus aliquid civibus nostris, prosimus etiam, si possumus, otiosi.

[Philosophy has lain neglected up to this day, nor has Latin literature shed any light on it. We must illuminate and revive it, so that, if I have done any good to my countrymen when I have been officially at work, I can do so when at leisure as well.]

Not only does Cicero cast himself as Archimedes' heir, but by overlapping the imagery used to describe the resurrection of philosophical inquiry, the invocation of Archimedes as an *exemplum*, and the discovery of Archimedes' tomb, he shows his own method of inquiry—one focused on ethics, relying ultimately on personal authority and the authority of the past rather than argument—as the heir and replacement of Greek theoretical inquiry.[50] Cicero is both less abstract and less concrete than Archimedes, who represents extremes of both abstract thought and the material expression of those ideas— *logos* and *ergon*, geometry and weaponry. As a philosopher, Cicero is practical and historical; as a defender of the state, he is political, an inventor of arguments rather than weapons. For Cicero, who is philosopher, rhetorician, and politician, ideas take shape as words, and words are deeds. Thus, even while embodying the polarity of *logos* and *ergon*, Cicero occupies a middle ground.

Other elements of the story lend themselves to an allegorical interpretation. With the Roman quaestor as their guide, the Syracusans leave their intellectually obsolete city to find their noble past outside its gates.[51] Cicero surveyed the field (*conlustrarem*) and approached the grave through the way cleared of brambles (might those *inmissi cum falcibus multi* represent previous philosophers?). There is no detailed description of the monument or quotation of the verse, no exclamation of "eureka!"—only the report of the report: *ego . . . dixi me illud ipsum arbitrari esse, quod quaererem* (I . . . said that I thought it was the very thing that I was seeking). One cannot approach the truth directly; one must fall back on the elements of rhetoric: digression, *exempla*, the *auctoritas* of the quaestor.[52]

Cicero has used Archimedes' tomb as a means of constructing this *auctoritas*. It helps him create a past. In a discussion of two of Cicero's dialogues, the *Laelius* and *Cato Maior*, Trevor Murphy has observed:

Time actively confers *auctoritas* on the events and the lives of the past. *Auctoritas* accumulates as one recedes into the past, back along a chain of stories repeated and disseminated. As memory lends past events and faces the power of accumulated emotion, so time confers on a bare action or event the status of *exemplum*. It is through time that men become not simply older but greater, *maiores*.[53]

Murphy goes on to discuss the historical settings of the *Laelius* and the *Cato Maior,* but what he says about time and authority holds for Cicero here as well. In this passage, Cicero generates *auctoritas* by fashioning a past Cicero. In a paradoxical process, the old man talks about the young in an old story embedded in a new dialogue. By embedding the old story, the narrating Cicero gives authority to the youthful Cicero, who, a discrete entity in narrative terms, becomes a forebear of the present Cicero, one of his intellectual *maiores.* This exemplary Cicero, older, greater, and an eyewitness participant in the crucial moment of (re)discovery, bestows his *auctoritas* on the narrating Cicero. Once again the *novus homo* is his own ancestor.[54]

The story of Archimedes' tomb, then, leads the reader toward the life of inquiry, a way of life that is now as Roman as it is Greek, thanks to Cicero's philosophical works, and one that is now as legitimate a pastime for a Roman as political service, thanks to Cicero's example of himself using his stint as public servant to satisfy his own curiosity.[55] It conveys its point by transmitting pleasure—Cicero's pleasure in inquiry and his pleasure in remembering his past life. The point of a digression is to please (*delectare*).[56] The very idea of a pleasure-causing rhetorical figure arises nicely from the context: Dionysius suspended a sword over the head of Damocles to convey vividly through a shared experience how fear of impending death could completely block the sensation of pleasure ("'so then, Damocles,' he said, 'since this life pleases [*delectat*] you, do you want to taste it yourself and experience my fortune directly?'" 5.62). Rather than portraying a scene that showed Archimedes happy and oblivious to the fear of death, Cicero conveys to his audience a sense of his own pleasure in inquiry through the intellectual pleasure of the digression. The anecdote achieves what the sword did in another way as well: the sword of Damocles blocks any thought of pleasure, even for a man sitting at a loaded table; the anecdote blocks any thought of death, even while describing a monument that is a reminder of a man's mortality.[57]

Cicero uses his past self as an exemplum for which he can vouch with complete authority. Returning to philosophy as a consolation, he remembers his past intellectual pleasure.[58] The sword hanging over Dionysius, Damocles, Archimedes, Caesar, and Cicero alike is replaced by the happiness common to all who live the life of inquiry. On the one hand, the technical arguments of the *Tusculan Disputations* aim to prove that the soul is immortal, that virtue suffices for happiness; on the other, the anecdote about the monument exhorts the reader to a virtuous (and, by implication, happy) life of inquiry, by presenting not just Archimedes as an exemplum but Archimedes and Cicero together.[59] In the continuous competition between the past and the present, the places associated with famous men urge the viewer to emulate them.[60] By including the anecdote in the final book of the *Tusculan Disputations,* Cicero shows that he has already successfully met this challenge, which his representation of the tomb reshapes and passes on to his audience.

Archimedes' tomb draws the reader toward the life of inquiry, just as the *figura* of cylinder and sphere standing out from the brush both illustrates the verses in a concrete way and beckons the person who is willing to search for truth that is not readily apparent. The rediscovery of the monument emerges as a symbolic point of intersection for the Greek and the Roman, for rhetoric and *doctrina,* for two ways of achieving the same thing: drawing a person away from obsession with the accidents of fortune in the sublunary world. At the same time, the monument holds the Greek and Roman worlds apart. The Greek life of inquiry belongs to the past, the Roman to the present and the future. The image of Cicero at Archimedes' tomb is Cicero's monument to his life of service and philosophy, and like other *monumenta,* it both reminds and exhorts. As Cicero, our exemplary Roman, represents it, the invitation to the life of inquiry is personal and specific: looking at Cicero's written monument, one wonders at human memory and, by considering the movements of memory, comes to contemplate the movements of the celestial spheres.

CHAPTER 3

WHY TWO SPHERES?

Cicero's account of finding Archimedes' tomb transforms the grave marker into a symbol of knowledge neglected and lost by its original owners, while the digression itself acts as an extended metaphor for the recovery and appropriation of that knowledge by a worthier heir. Written late in Cicero's career and looking back to one of that career's early landmarks, *Tusculan Disputations* 5.64–66 incorporates Archimedes, in his dual role as thinker and defender of the state, into Cicero's intellectual and political autobiography. In addition, by recalling the reference to Archimedes' sphere in the first book of the *Tusculan Disputations,* the grave marker helps unite the dialogue's initial argument for the immortality of the soul and its later claim that *virtus* suffices for living happily. Finally, that Cicero makes so much of his memory of an event in Syracuse and does so within the greater discursive framework of the Roman appropriation of Greek culture points to the city's significance as a site both of his own inspiration and of disruptions in the continuity of even the most accomplished of cultures.

As we have seen, the structure of this digression is a crucial part of the image it conveys, which is not of the tomb itself but of Cicero's act of discovery. This structure links the image of discovery to other contexts external to the *Tusculan Disputations,* those of politics and memory, in which discovery also entails a return to a point of departure. Cicero's journey to the tomb links him to the one other Roman mentioned in connection with it, Marcus Claudius Marcellus, conquerer of Syracuse, who saw to Archimedes' burial.[1] In raising Archimedes' spirit, then, Cicero summons up that of the philhellenic Marcellus as well. As for memory, Cicero would not even have found the tomb had he not remembered the verses relating the relative volumes of sphere and cylinder. Of course, embedded narratives exclude as well as include: this passage shies away from referring directly to Archimedes' death; it mentions neither the capture of Syracuse, nor the years of Roman rule that may have contributed to the tomb's neglect. Even the remembered Greek *senarioli,* unquoted, are lost to us.

Cicero's description of Archimedes' tomb involves yet another return, to themes of the earlier *De republica,* where two spheres attributed to Archimedes—one solid, one orrery—serve as extended metaphors that make programmatic statements.[2] The embedding of the spheres in the *De republica* is more complex than that of the tomb in the *Tusculan Disputations,* because the *De republica* is more truly a dialogue and brings more voices into play.[3] Yet, much as the tomb digression links the ideas of the *Tusculan Disputations* to Cicero's life of politics and inquiry, so, too, the representation of the spheres links major concerns of the *De republica* to Cicero's personal experience.[4] Cicero wrote the *De republica* in 54–51 BCE but chose to set it early in 129 BCE, shortly before its main figure, Scipio Africanus, died in mysterious circumstances.[5] Giving the dialogue a setting more than twenty years before his own birth meant that Cicero could make autobiographical references only in the framing material, the prefaces addressed to his brother Quintus. Moreover, some of this material—indeed, much of the dialogue—is lost, a circumstance limiting what we can say conclusively about the work as a whole.[6] Still, we can see from what remains that Cicero's discussion of the spheres places emphasis on memory, discovery, and the transmission of knowledge; moreover, the manner in which the spheres are embedded in the *De republica* binds together their several frameworks—discursive, physical, social, and autobiographical—with the result, to paraphrase Emma Gee, that the organization of the narrative mirrors the design of the world.[7] Here, as in the *Tusculan Disputations,* the way in which the exemplum is embedded also constitutes a crucial part of its message.

The Setting of the Dialogue and the Description of the Spheres

Early in the surviving fragment of book 1 and, in the dramatic setting of the dialogue, early on the first day of the Latin Festival, before the participants in the discussion have assembled in Scipio Africanus's gardens, Scipio's nephew, Quintus Tubero, asks his uncle about a phenomenon recently reported to the Senate, the appearance of a second sun (1.15).[8] His uncle only begins to reply:

after saying that he wished Panaetius were present to give an informed opinion on the phenomenon, then touching on Socrates' preference for ethics
over natural philosophy and summarizing Plato's intellectual development
(1.15–16), Scipio spies the newly arrived Lucius Furius Philus and greets him,
together with Rutilius Rufus, who has arrived at the same time. Philus asks
Scipio and Tubero what they are talking about. Scipio says it is a topic that he
and Rufus used to discuss under the walls at Numantia—namely, the two
suns. This is a topic of interest to Philus, and Scipio is keen to hear what
Philus has to say about it (1.17).

Before Philus can respond, a slave announces Laelius's imminent arrival.
With Laelius are Spurius Mummius and Laelius's own sons-in-law, Caius
Fannius and Quintus Scaevola. All present walk about, then take seats in a
sunny spot for conversation. Manius Manilius arrives and sits down next to
Laelius (1.18). Philus says that the group should continue its conversation.
Laelius asks what conversation has been interrupted. When Philus tells him it
was talk of the two suns, Laelius asks if they are so sure that they have such
complete knowledge of matters of home and state that they can spend time
discussing celestial phenomena. Philus replies that "home" should include
the whole universe; besides, he adds, such learning gives him pleasure, as it
surely does Laelius and others who are greedy for knowledge (1.19). Are they
too late to hear what Philus has to say, asks Laelius. No, replies Philus, but
perhaps they might hear Laelius's opinion. Laelius, however, would rather lis-
ten to Philus, unless, he adds, Manilius wants to pronounce an edict that the
two suns should reach a compromise. But Manilius, too, would rather hear
what Philus has to say than speak himself (1.20). So Philus complies.

I will return to this preliminary material; for now, let us move on to
Philus's words. I quote them in full, because what follows discusses the passage in some detail. Philus begins with his memory of seeing, in Marcus Marcellus's house, a sphere made by Archimedes, which he compared to another
that was in the temple of Virtus. The one in the temple was a solid sphere, the
other was an orrery. Both, it seems, were brought to Rome by Marcellus's
grandfather after the sack of Syracuse.[9] I have marked references to the solid
sphere with (1) and references to the other with (2).

> Tum Philus: 'Nihil novi vobis adferam, neque quod a me sit cogitatum
> aut inventum. Nam memoria teneo, Gaium Sulpicium Galum, doctis-
> simum ut scitis hominem, cum idem hoc visum diceretur et esset casu
> apud Marcum Marcellum qui cum eo consul fuerat, sphaeram (2)
> quam Marci Marcelli avus captis Syracusis ex urbe locupletissima

atque ornatissima sustulisset, cum aliud nihil ex tanta praeda domum
suam deportavisset, iussisse proferri; cuius ego sphaerae (2) cum per-
saepe propter Archimedi gloriam nomen audissem speciem ipsam
non sum tantopere admiratus; erat enim illa (1) venustior et nobilior
in volgus, quam ab eodem Archimede factam posuerat in templo Virtutis Marcellus idem; sed posteaquam coepit rationem huius operis
(2) scientissime Galus exponere, plus in illo Siculo ingenii quam.
videretur natura humana ferre potuisse, iudicavi fuisse. Dicebat enim
Galus sphaerae illius alterius solidae atque plenae (1) vetus esse inventum, et eam a Thalete Milesio primum esse tornatam; post autem ab
Eudoxo Cnidio, discipulo ut ferebat Platonis, eandem illam astris eis
quae caelo inhaererent esse descriptam; cuius omnem ornatum et
descriptionem, sumptam ab Eudoxo, multis annis post non astrologiae scientia sed poetica quadam facultate versibus Aratum extulisse;
hoc autem sphaerae genus (2), in quo solis et lunae motus inessent, et
earum quinque stellarum quae errantes et quasi vagae nominarentur,
in illa sphaera solida (1) non potuisse finiri; atque in eo admirandum
esse inventum Archimedi, quod excogitasset quemadmodum in dis-
simillimis motibus inaequabiles et varios cursus servaret una conversio. Hanc sphaeram (2) Galus cum moveret, fiebat ut soli luna totidem
conversionibus in aere illo, quot diebus in ipso caelo succederet, ex
quo et in caelo <et in> sphaera solis fieret eadem illa defectio, et in-
cideret luna tum in eam metam quae esset umbra terrae, cum sol e
regione [***]' (1.21–22).[10]

*[Then Philus replied: "I shall bring before you nothing that is new nor
anything thought out or discovered by me. For I remember this of C.
Sulpicius Gallus, as you know a most learned man: when this same thing
[two suns] was being said to have appeared, and he was, by chance, at the
house of Marcus Marcellus, who had been his colleague as consul, he commanded that the sphere be brought forth (2), which Marcellus's grandfather, having captured Syracuse, had carried off from that very wealthy
and most magnificent city, although he had taken home nothing else from
such great booty; and the appearance of this sphere, though I had quite
often heard mention of it on account of Archimedes' fame, I did not ad-
mire so much. For that other one (1) was more charming and more commonly known, which was made by the same Archimedes and which that
same Marcellus had placed in the temple of Valour. But after Gallus began
most knowledgeably to explain the system of this work (2), I decided that*

there was more genius in that Sicilian than human nature seemed able to admit of. For Gallus was saying that the discovery of that other, solid and filled sphere (1) was an old thing, and that it had been first turned by Thales of Miletus, and that afterward that same sphere had been inscribed with the stars that cling to heaven by Eudoxus of Cnidos, a pupil, as he was saying, of Plato—all of which ornament and description, Aratus had taken from Eudoxus many years later and had extolled in verse, not with astrological knowledge, but with a certain poetic skill. This kind of sphere (2), however, in which were represented the movements of the sun and the moon and of those five stars that wander and are called, as it were, errant, could not have been delineated in that solid sphere (1); and in this respect the discovery of Archimedes was all the more to be marveled at, that he had thought out how one revolution might preserve the unequal and varied courses with their very dissimilar movements. When Gallus set this sphere (2) in motion, it came about that the moon went under the sun with the same number of turns on that bronze as days in the heavens themselves, so that in the sphere too there came about the same eclipse of the sun as in the heavens, and the moon came to the turning point where the shadow of the earth was, at that time when the sun . . . from the region. . . ."][11]

The manuscript breaks off as Philus is using the second, mechanical sphere to explain an eclipse. A gap of two (or possibly six) manuscript pages follows.[12] Gallus may have concluded his demonstration by pointing out the structure of the heavens, from which Philus would have argued that the two suns were simply a deceptive reflection, not a prodigy.[13]

The loss of Philus's explanation has not kept scholars from pointing out that this passage has thematic connections with the rest of the dialogue: Emma Gee has shown how the vocabulary describing the first sphere conflates text and object; Robert Gallagher has explored the ways in which the vocabulary describing the innovative sphere anticipates Scipio's talk of the cycle of constitutions and his dream in book 6, so that the image of the sphere unites the dialogue's beginning and end.[14] While the importance of the passage to the later part of the dialogue has been well established, we still need to consider the spheres in the context of its beginning. The remaining question, perhaps too obvious, is not Tubero's "Why two suns?" but "Why two spheres?" To answer it, we need first to examine the description of Archimedes' spheres and then to consider a pair of related topics: (1) what the passage says about the spheres' provenance and (2) how the passage both draws meaning from the context that frames it and, in turn, influences the meaning of that context.

Comparison

Philus introduces the older, solid sphere by way of comparison: to account for his initial lack of wonder at the appearance of this odd planetarium, he points out that the solid sphere is "more charming" (*venustior*) and "more known" (*nobilior*) to the general public. The very structure of the passage encourages readers to compare the two spheres, for Philus's description alternates between them, beginning with the second, newer one.[15] This alternation, marked by various forms of the demonstratives *hic* and *ille*, turns readers' attention from one sphere to the other, thus generating an image of the two side by side. Philus says, moreover, that the older, solid sphere was made by the same Archimedes who produced the mechanical one (*ab eodem Archimede factam*) and that it was placed in the temple of Virtus by the same Marcellus who kept the new planetarium at home (*Marcellus idem*). The solid sphere is an old discovery (*vetus esse inventum*), reproduced by Archimedes, whereas the mechanical sphere is Archimedes' own (*inventum Archimedi*). The facts that both spheres have the same maker and that the same person brought them to Rome clearly invite exploration of the spheres' differences. Yet since the newer, mechanical sphere is the one used by Gallus to illustrate celestial phenomena, the reference to the older, solid sphere seems gratuitous. Why is it here at all? Why does Philus juxtapose old and new in a way that appears to introduce the old solely to privilege it as more charming and better known?

In fact, conjuring up the newer sphere in a way that juxtaposes it and the older one has the paradoxical effect of making the older one the lesser of the two, although it is the more charming and better known.[16] Philus introduces the old sphere only as a comparandum for the new; he also asserts the newer sphere's superiority by saying that it was when Gallus explained its system, that he, Philus, made his own judgment (*iudicavi*) about Archimedes' genius being greater "than human nature seemed able to admit of." I will return to the effects of this comparison; for now, let us consider the effect of another apparently unnecessary element of the description, Gallus's account of the history of both spheres, who made them, how they came to Rome, and where Marcellus put them.

Provenance

Spheres are portable; and the Latin places emphasis on that quality: Marcellus had removed Archimedes' sphere from Syracuse (*sustulisset/deportavisset*); he had placed (*posuerat*) the solid one in the temple of Virtus; Gallus ordered that the other, the one kept at home, be brought out (*iussisse pro-ferri*). Even the account of the older sphere's literary history employs metaphors of taking and removing: Aratus took the plan of the sphere from Eudoxus (the verb is *sumo*) and described it in his verse (the verb is *effero*).[17] Since spheres are portable, they make convenient booty; and it is as booty that these came to Rome (*captis Syracusis/ex tanta praeda*).[18] Since a sphere represents the cosmos, one can readily agree with scholars who interpret the arrival of Archimedes' spheres in Rome as representing the transfer of world dominion to its new possessor.[19] Yet both the very act of telling their stories while describing them and the reference to the beauty of the older sphere make the nature of these prestige items more complex, for details of their de-scription recall the prizes and guest-gifts of the Homeric poems.

Consider, for example, the mixing bowl presented at Patroklos's funeral games in *Iliad* 23.740–49 (trans. Lattimore).

> At once the son of Peleus set out prizes for the foot-race:
> a mixing-bowl of silver, a work of art, which held only
> six measures, but for its loveliness it surpassed all others
> on earth by far, since skilled Sidonian craftsmen had wrought it
> well, and Phoenicians carried it over the misty face of the water
> and set it in the harbour, and gave it for a present to Thoas.
> Euneos, son of Jason, gave it to the hero Patroklos
> to buy Lykaon, Priam's son, out of slavery, and now
> Achilles made it a prize in memory of his companions.

Perhaps even more pertinent is the talk accompanying the gift from Menelaus to the departing Telemachos in *Odyssey* 15.113–19 (trans. Lattimore).

> Of all those gifts that lie stored away in my house I will give you
> the one which is most splendid and esteemed at the highest value.
> I will give you a fashioned mixing bowl. It is of silver,
> all but the edges, and these are finished in gold. This is
> the work of Hephaistos. The hero Phaedimos, the Sidonians'
> king, gave it to me, when his house took me in and sheltered me
> there, on my way home. I would give it to you for a present.

Homeric mixing bowls, like spheres, are portable: they can be carried over the water, placed in the harbor, set out as prizes, or given to departing guest-friends. They have other common features: both have illustrious inventors or makers (Sidonian craftsmen and Hephaistos/Thales and Archimedes);[20] they are beautiful and esteemed; they are imported across the water (from Phoenicia/from Syracuse); they lie stored away in a house (Menelaus's/Marcellus's), ready to be brought forth at the right moment and presented to a promising and well-connected member of the next generation (Telemachus/Philus). Like the sphere, moreover, the Homeric gift or prize is select, the most excellent of its kind: "for its loveliness it surpassed all others / on earth by far" (αὐτὰρ κάλλει ἐνίκα πᾶσαν ἐπ᾽ αἶαν / πολλόν). "Of all those gifts that lie stored away," says Menelaus, he gives Telemachus the one "most splendid and esteemed at the highest value" (κάλλιστον καὶ τιμήεστατον).

The parallels between these Homeric prizes and Ciceronian spheres, however, draw attention to a some striking differences. First, the spheres are identified as prizes of war, the new planetarium explicitly so. Second, "from such great booty" (*ex tanta praeda*), "from that very wealthy and most magnificent city" (*ex urbe locupletissima atque ornatissima*), Marcellus chose one piece to keep at home, and that one, select item is brought forth; but to Philus's eyes, the mechanical sphere by itself merits no Homeric superlatives. It does not even fit into the conventional categories of booty, being neither a weapon taken from an enemy to adorn the aristocratic Roman house nor an ornament, like the statues and paintings in the temples of Honos and Virtus.[21] Had Marcellus wanted an *ornamentum* for his house, he could have kept the older, solid sphere, which is the more charming of the two and thus alone has the potential to be described in Homeric terms as the "most splendid."

Yet even as Philus explains his initial lack of admiration for the mechanical sphere, the text draws attention to a shift in values. The sphere more charming to the eye gives way to the one that, when interpreted, evokes awe, not for the sphere itself, but for the talent of its absent maker. At the same time, the superlatives in the passage, which are never applied to the spheres themselves, move from Syracuse (the richest and most decked-out city, a prize in its entirety) to Gallus, who explains most knowledgeably (*scientissime*). The Latin thus registers a shift of interest from riches and ornament to knowledge and understanding.[22]

A shift in the value of provenance parallels this shift in interest. The value of the heroic prize lies partly in its craftsmanship and partly in its provenance—who created it, the story of who created it, who owned it first, who owned it next, who carried it over the water, that it *was* carried over the water.[23] Gallus traces the development of the older sphere from its inventor, Thales, through Eudoxus and Aratus, to Archimedes, who reproduced it. Philus, too, participates, tracing its history beyond Archimedes to Marcellus, who carried it across the water and placed it in the temple of Virtus. In Philus's description, the older sphere acquires the significant features of its provenance over a long period of time and before its arrival in Rome. Note the temporal references: "first . . . , afterward . . . , many years later" (*primum . . . , post . . . , multis annis post*). The second sphere, in contrast, seems not to have acquired the prestigious patina of the first: no Thales; no Eudoxus, pupil of Plato; no Aratus. But in Rome its story continues: Gallus ordered it brought forth; he explained its system and set it in motion. Note, again, the temporal reference: it was, says Philus, "after [*posteaquam*] Gallus began to set out the system" of the sphere that it evoked his admiration. Philus himself remembers the sphere's movements and, when the manuscript breaks off, is describing them in detail to an audience that includes Rufus, who remembers Philus's story and passes it on to a youthful Cicero, who in turn remembers Rufus's report and records his memory for himself, his brother, and his other readers.

The first sphere, then, has an illustrious provenance, beginning with its discovery by Thales, first of the Seven Sages, and including the contributions of a series of learned men. This provenance links the first sphere to various discourses (Eudoxus's *Phaenomena* and Aratus's poem on the same topic, also entitled *Phaenomena*), and it is mostly Greek—Milesian, Cnidian, and Sicilian, with a little Roman contributed at the end by Marcellus. In fact, since this first sphere is Archimedes' imitative reproduction, what has been passed down from Thales is not the artifact fashioned by the sage's very hands but the knowledge, in the form of a treatise or model, of how to make such a sphere.[24]

The second sphere, in contrast, is no reproduction; its story begins with its invention by Archimedes, and that story has been Roman ever since. The discourses associated with it are neither Eudoxus's treatise nor Aratus's poem (nor even Cicero's translation of Aratus) but Gallus's explanation and Philus's representation—the layers of remembered conversations that make up the *De republica*; for Philus remembers not the spheres themselves but his reaction to them and Gallus's ordering the mechanical sphere brought forth. The image of that second sphere is, in fact, incomplete until Philus—after he recollects Gallus telling the history of both spheres, Gallus setting the second sphere in motion, and Gallus using it to explain the planetary system—finally takes over himself and describes its workings in detail.[25] It is, then, as a result of Gallus's explanation and Philus's story that the second sphere acquires layers of significant associations.

Thus, in addition to the contrasts between old and new, public and private, common and arcane, embodied in these two spheres, we find another:

the contrast between that which is visually charming and that which relies on verbal explanation. While the older sphere, which draws the admiration of the masses, is a choice piece of booty from the city that exemplified rich spoils, the newer sphere, described with an emphasis on selection and exclusivity, knowledge and understanding, becomes a prestige item of another sort.[26] The first sphere charms viewers without requiring them to comprehend anything. The second sphere requires explanation and setting in motion, that is, the presence of a knowledgeable interpreter. Interpreting it involves uniting not only words and illustration but also words and deed, a combination of *logos* and *ergon*. Thus the second sphere, once set in motion, is a fitting emblem for the dramatic philosophical dialogue, which also represents the words and deeds of its participants and illustrates its author's ideas dynamically.[27] The job of the second sphere, after all, is to illustrate the spatial and temporal relationships of objects in motion. If the second sphere can be said to embody arcane knowledge, we can say that Gallus's dynamic interpretation of it represents the transmission of that knowledge through dialogue, which links the elite of one generation and the next.

Such a role fits in nicely with the second sphere's provenance, most of which is generated within the dialogue itself; it is also appropriate to the setting of the explanation in a private house and the select nature of the participants. (Karl Büchner says that the comparison of spheres symbolizes the way in which education elevates the intellectual elite above the masses.)[28] A circle of intellectual aristocrats can thus transform the process of appropriation from the collection of beautiful artifacts into the interpretation and timely transmission of cultural capital, the arcane knowledge illustrated by an artifact. While the transfer of the spheres to Rome symbolizes the transfer of power, the fact that there are two also reminds readers that there are varying ways to take possession of the world's riches and, indeed, different ways of defining and valuing those riches. I will revisit the idea of the spheres as illustrations, for in addition to representing the dynamics of dialogue, they also illustrate other cultural movements.

Continuity and Change

Cicero identifies the circumstances of this fictional conversation very precisely: during the Latin Festival (*feriae Latinae*), in the gardens (*in hortis*) of P. Cornelius Scipio Africanus, when Tuditanus and Aquilius were consuls.[29] The date is some fifty years before Cicero and his young brother heard Rutilius Rufus's account of the dialogue, at Smyrna, and over seventy years before

Cicero began writing the *De republica*. The historicizing details help to emphasize the challenges faced by those who want to preserve knowledge: expanses of time and distance and the replacement of one generation by the next. Yet the dialogue also draws attention to time, place, and persons as part of a system of memory and its transmission, a system that brings about both cultural continuity and fruitful change. James Zetzel has pointed out that what strikes the reader of the *De republica* is its display of tradition and knowledge passed down personally.[30] Scholars have noted the cultural continuity embodied in the sets of older and younger participants in the dialogue: each of the younger set has a close relationship to either Scipio or Laelius.[31]

Philus's account of the spheres continues to place emphasis on this continuity. The setting in which he sees the second sphere, the house of the Marcelli, links Marcellus *avus* and his grandson. A republican institution, the consulship, links the younger Marcellus to Gallus. The explanation of the spheres links Gallus and Philus.[32] Gallus, then, brings forth what was tucked away in a manner reminiscent of a Homeric hero's bringing forth, at the appropriate time, a guest-gift that forges ties across generations. Menelaus, when presenting the bowl to Telemachus, recollects the person from whom and circumstances in which he received it. So, too, Philus, when passing on his story, recollects the person with whom, the place where, and the circumstances under which he saw Archimedes' sphere: Gallus; Marcellus's house; and the previous instance of the second sun. Just as the Homeric gift creates a bond between giver and receiver, the story of this second sphere creates a bond between Gallus, who explains and transmits arcane knowledge, and Philus, who receives and understands the explanation. By remembering the explanation, bringing it forth at the appropriate time, and transmitting it in turn, Philus creates a bond with his own audience.[33]

Like Menelaus recollecting the history of the bowl and Philus recollecting the story of the sphere, Cicero recollects for his brother the person from whom and the circumstances in which they heard the dialogue.

> Nec vero nostra quaedam est instituenda nova et a nobis inventa ratio, sed unius aetatis clarissimorum ac sapientissimorum nostrae civitatis virorum disputatio repetenda memoria est, quae mihi tibique quondam adulescentulo est a Publio Rutilio Rufo, Smyrnae cum simul essemus compluris dies, exposita. (1.13)

> *[And truly I am not about to state principles that are new or were discovered by me, but it is my intention to recall a discussion carried on by men who were at a one period the most eminent and wisest in our state. This*

*discussion was once reported to me and to you in your youth by Publius
Rutilius Rufus, when we were together at Smyrna for several days.]*

According to Cicero, this dialogue was a highly mediated experience, old and
to be approached through memory. He is only now setting down recollec-
tions from when he was younger and when Quintus was still an *adulescentu-
lus.* Moreover, Cicero and Quintus were not even born when the original discussion took place. They heard it secondhand.[34]

What strikes the reader working through the preface is how Philus's introduction of the two spheres echoes Cicero's introduction to the dialogue
or, to follow the chronology of Cicero's fiction, how Cicero models his approach to the dialogue on his memory of what Rutilius said that Philus said
about the spheres. What Cicero would write was not new (*Nec . . . nova*) nor
discovered by him (*a nobis inventa*). Philus's memory of Gallus is something
not new (*Nihil novi*) nor thought out or discovered by him (*neque quod a me
sit cogitatum aut inventum*).[35] Cicero says that he must use memory (*sed . . .
repetenda memoria est*); Philus uses memory to grasp what Gallus told him
(*memoria teneo*). Rutilius Rufus set out the dialogue for Cicero and his
brother (*exposita*); Philus uses the same verb of Gallus explaining the *ratio* of
the sphere (*exponere*). Even the superlatives in Cicero's introduction—the
men involved in this *disputatio* were the most illustrious (*clarissimorum*) and
wisest (*sapientissimorum*) of their time—are matched by those in Philus's
description—the sphere was booty from the richest (*locupletissima*) and
most magnificent (*ornatissima*) city of all. Finally, as we track the relationship of the sphere and the frame narrative, it is worth noting that Philus and
Rufus, the one the source for the account of the spheres and the other the
source for the account of the entire dialogue, arrive at Scipio's gardens together and that the text emphasizes this fact.[36] Each layer of embedding thus
both contributes to the spheres' provenance and links Cicero and his audience to Archimedes and Thales. By saying that he will bring in nothing that is
new or that he has thought out or discovered, Philus places himself at the end
of a tradition that he merely reports. Noting, however, the resemblance between his words and Cicero's, we can also say that Philus is located at the
midpoint of a tradition that Rutilius Rufus and then Cicero report. It seems
that in transforming himself—and the works of his predecessors—Cicero
presents himself as simply following tradition. The "new man" is not doing
anything new; in transmitting a vision of a remembered world and vividly
representing the artifacts of Archimedes, he is simply displaying the *monumenta* of his intellectual ancestors.[37]

Yet embedded deeply within this framework that so ostentatiously privileges the old is the vivid image of the second sphere evoking Philus's awe as something new. Moreover, while the first sphere is charming, this second, new sphere is, in Gallus's hands, *utilis*.[38] When, after the lost folio pages, the manuscript picks up again, Scipio praises not Archimedes' ability to invent a sphere nor Gallus's display of knowledge about it but, rather, Gallus's ability to calm panicked troops and affect the outcome of a battle by using that knowledge rhetorically (1.23). The choice piece of foreign booty has gained a Roman provenance, has become a part of Roman discourse in terms of its usefulness for Romans.[39] The old discovery, the more charming, concrete, and particular solid sphere, emerges from this passage inadequate. Judgment falls on the side of the new and useful with the result that the arcane, abstract, and, surprisingly enough, more theoretical becomes, in the hands of Gallus and then Scipio, the more useful.[40] In Philus's act of judging (*iudicavi*), located so precisely in time and space, we see Cicero's representation of the point in time on which this shift in perspective turns. Archimedes' spheres stand at the focal point of the new perspective on the past.[41]

As a contrasted pair, the two spheres complicate the sense of dualism running through the work as a whole (old/new, Greek/Roman, theoretical/practical, abstract/concrete).[42] The comparison of the two spheres sets up its own series of contrasts that seems to run counter to these: contrasts between vulgar knowledge and arcane, remembered and discovered, which, at one level of the narrative, appear to privilege the old (Cicero and Philus insist that what they have is not new; the older sphere was "more charming," says Philus), but which, in the deeply embedded description of the spheres, turns out to champion the new. When Philus considers the spheres as technological inventions, the newer makes the old the weaker of the pair. This passage, then, shows the intersection of two value systems: one heroic, in which the acquisition of provenance improves and adds value to the new by giving it tradition and prestige; the other technological, in which discovery (*inventio*) improves and adds value to the old.[43] This juxtaposition of old and new technology represents a shift in the perspective from which one views the past tradition, for the passage reminds readers that although the past is valued, the present sets the terms by which to evaluate the past. Its Greek and Syracusan provenance is part of the value of the old, solid sphere now in Roman hands; but now we see the Roman tradition adding a different kind of value to the sphere invented in Syracuse.

Another, related double image arises from this description of the spheres. When Tubero suggests discussing the two suns, Scipio wishes that Panaetius, who was interested in this sort of thing, were present, although, adds Scipio, Socrates seemed wiser for refusing to inquire into such matters. Tubero says that he cannot understand how the tradition started that Socrates was interested only in human life and morals. He points out that Socrates combined the study of ethics and politics with that of arithmetic, geometry, and harmony, in the manner of Pythagoras. Scipio answers that one needs to make a distinction between Socrates and Plato, for after Socrates died, Plato traveled to Egypt, Italy, and Sicily, where he became acquainted with Pythagoras's teachings.[44] Scipio presents Tubero with a biography of a twofold Plato, who transmitted and transformed Socrates' visions; Philus lays emphasis on the fact that the same Archimedes produced two spheres—one that looked backward, one that was new—and that the same Marcellus, victorious general and philhellene, displayed one sphere and kept another hidden away.

The use of the second sphere as an illustration recalls a comment in Plato's *Timaeus* about the futility of describing the movements of the heavenly bodies without a visible representation of them.[45] Such an illustration, according to the *De republica* passage, was not available in Plato's time; for even the older sphere first "turned" by Thales was elaborated in the post-Platonic world. It was first inscribed with stars by Plato's pupil Eudoxus, then extolled in verse by Aratus; it took the post-Platonic Archimedes to illustrate the movement of heavenly bodies. Thus the second sphere nicely underscores the dialogue's strong interest in post-Platonic "working" models. In doing so, moreover, it suggests Polybian inspiration, if not Polybius as a direct source. We do not know if Polybius, who described Archimedes' defense of Syracuse so carefully, ever described a sphere of Archimedes; but he did describe a system in motion: the movement of Fortune and the workings of history. He insisted, too, that one must see the whole of history and that one must write history from experience, for not to do so is, as it were, to paint pictures of animals from the parts of dead ones, not the living moving reality.[46] The two spheres, then, represent nicely two ways of looking at the Roman past.[47] On the one hand, there is the recording of *prodigia* without explanation that forms the kernel of the annalistic tradition. This is represented by the first sphere, a work of art set up in a temple and not requiring interpretation. On the other, there is Polybian history, which, like the second sphere, takes a comprehensive view of systems in motion and emphasizes rationalism, chronological accuracy, political theory, and the practical. Indeed, Polybius discusses at length (and with examples) the advantages to the general of knowing astronomy.[48] We have, then, in the *De republica*, a collection of visible working models: Archimedes' working model of a sphere; Polybius's working model of history; the Romans' working model of a *res publica*.[49] By bringing the two spheres into his *De republica* in a way that refers to the *Timaeus* (and anticipates his own translation of the work), Cicero draws attention to the time that has passed and the intellectual progress made since Plato and his *Republic*.[50] The second sphere thus provides material evidence that just as there were new ideas in the post-Socratic world of Plato's *Republic,* so, too, there are new ideas in the post-Platonic world of Cicero's *De republica*.[51]

Thus the spheres illustrate the intellectual pedigree that Cicero claims for his third- and second-century Romans and also for himself. One sphere leads all the way back to the greatest of the Seven Sages, the other back to Archimedes and his invention.[52] The connections forged by the framing narratives bring to light yet another set of connections, this one between Plato, Archimedes, Marcellus, and Cicero, all of whom spent time in Sicily. Plato studied with Socrates, traveled in Egypt after Socrates died, then went to Italy and Sicily, where he learned new ideas from Pythagoras. The Sicilian Archimedes (Philus calls him "that Sicilian" [*illo Siculo,* 1.22]) made his first sphere by imitating the old ideas of Thales and then moved on to invent the new and improved. Marcellus, himself a man of the Greek world as well as the Roman, one who both conquered Syracuse and became patron of the Sicilians, saw to it that those spheres came to Rome. As we have seen in the *Tusculan Disputations,* Cicero represents his own stay in Sicily as important to his political and intellectual life. It is as if Cicero's mythologizing replaces the sojourn in Egypt—so formative for Plato, for Thales and Pythagoras, and even for Archimedes himself—with the stay in Sicily, which, like Athens and Alexandria, has now given way to Rome.[53]

Transmission and Exposure

Reading the opening passages as a series of layers of provenance that alter the relative value of Archimedes' two spheres may seem an odd means by which to approach the *De republica.* But Philus's act of describing unites provenance and the dramatic setting, the latter of which, as Zetzel points out, is very much a part of this picture.[54] Let us examine this setting more closely, returning for a moment to the opening discussion and the phenomenon of the two suns.

The striking features of this discussion are, first, that it represents talk about talk and, second, that the talk itself, taken up anew each time a participant arrives and joins the conversation, draws attention to each act of interruption. Throughout the preliminary discussion (1.14–20), the participants discuss not the two suns but the idea of discussing such a phenomenon. No one present claims actually to have seen the two suns. Tubero says to his uncle:

> Are you willing, then, since you, in a way, invite me and give me hope of hearing your answer, that we should look to this first, Africanus, before the others arrive: what it is that has been announced to the Senate [*quod nuntiatum est in senatu*] about the second sun? For they are neither few nor trifling who say [*qui . . . dicant*] that they have seen two suns; therefore it is not so much a matter of believing them as of inquiring the reason. (1.15)[55]

Subsequent references to the suns are also couched in terms of the talk about them: "'Surely our arrival has not broken up some conversation of yours [*sermonem vestrum aliquem*]. . . . What was the particular subject?' . . . '[It was] about those two suns [*de solibus istis duobus*]'"; "'What conversation have we interrupted [*cui sermoni intervenimus*]?' 'Scipio had asked me [*quaesierat ex me Scipio*] what I thought of the agreed-upon fact that two suns were seen.'" Even Laelius's legal joke enshrines the suns in a particular kind of talk, the language of law: "on the contrary, let us hear what you have to say [*te audiamus*], unless by chance Manilius thinks that a compromise should be reached, that they possess the heavens in such a way that each possesses it [*nisi forte Manilius interdictum aliquod inter duos soles putat esse componendum ut ita caelum possideant ut uterque possiderit*]." Functioning purely at the level of discourse, the two suns thus initiate discussion, spur debate, and prompt memory.

The interruptions in this initial conversation are strongly marked. Tubero tries to avoid interruption by asking for the explanation "before the others arrive" (*ante quam veniunt alii*). Furius is concerned that their arrival (*interventus*) has broken up (*diremit*) the conversation. Laelius too, asks, "What conversation have we interrupted?" (*cui sermoni nos intervenimus?*). Interruptions can ruin conversations, but in this case they expand a private talk between an uncle and nephew into a discourse on the commonwealth between several distinguished men of two generations; thus the interruptions generate the dialogue, each one pushing the talk along and altering its course, guiding it toward Philus's description of the spheres and on to Scipio's discussion of the utility of such knowledge and his discourse on kingship.

Like this preliminary discussion, the story of knowledge illustrated by the spheres is one of breaks and alterations in course; and these interruptions can be seen, from Cicero's Romanocentric perspective, as beneficial to that knowledge. Philus represents not so much an ekphrasis of the spheres as his own memory of Gallus's explanation and his own act of judging. The sphere invented by Thales was inscribed with constellations by Eudoxus and translated into another form entirely by Aratus, who versified it, "not with astrological knowledge, but with a certain poetic skill" (*non astrologiae scientia sed poetica quadam facultate*). Marcellus sacked Syracuse and brought the famous booty back to Rome. The act was criticized by Polybius, as it would be later by Livy; but Cicero, speaking elsewhere and for other purposes, claimed that Marcellus refounded the city when he captured it; and a character in Livy, speaking from an imperialist and Romanocentric point of view, would later claim that the Romans freed Syracuse by capturing it.[56] A Romanocentric version of the sack of Syracuse could claim that Marcellus took that second sphere under his protective care until someone could set it out at the appropriate time as a valuable gift/prize for the next generation.[57] After all, the Syracusans, who lost track of Archimedes' grave, would surely have been irresponsible custodians of the spheres.

If Archimedes' spheres are simply Homeric prizes to be handed down from individual to individual, then Rome's gain comes at the expense of Syracuse. Cicero, however, breaks the rules of this zero-sum game: the first sphere is available for public viewing; more important, the second sphere has now been revealed, and this revelation involves more than simply handing down knowledge, as if it, too, were an artifact. What is most significant here is not just that the second sphere is incomplete without someone to set it in motion or that Gallus's explanation of it is futile unless Philus understands it, remembers it, and passes it on. Nor is it simply that Archimedes' genius is now dependent on Marcellus, the philhellenic looter; Gallus, the interpreter; Philus, who understands and remembers; Rutilius Rufus, who remembers; and Cicero, who remembers and records. I have been arguing that the description of the spheres expands the idea of appropriation and implies that while one can physically remove another culture's artifacts, one can also reshape its representations or translate and transform its texts; in doing so, one can transform and transmit its knowledge. By doing so, moreover, one adds to the provenance of both the object and text and thus to its value. As the *De republica* presents it, each person who passes on the story of the sphere adds value; what makes this increase in value possible for the second sphere is a violent interruption, its capture at the sack of Syracuse.[58]

Archimedes left permanent artifacts embodying his visions of the cosmos. For a time, Marcellus controlled their fate. Philus describes the transmission of this vision from generation to generation, by a person setting the sphere in motion and explaining its movements.[59] As Cicero represents it, another vision of the cosmos, the one transmitted by Africanus to his grandson (his adoptive grandson—Scipio's two fathers are another instance of the dialogue's "double vision") and entrusted to his memory, is also transmitted orally person to person, until Cicero writes it down.[60] Technological interruption, the transition from oral explanation to written, is what makes it possible to restore the spheres to Syracuse, because it expands the number of people who can come into contact with them by reading the *De republica*. In written form, accounts of the knowledge conveyed by the spheres, knowledge made useful in Roman terms, can travel simultaneously anywhere, even back to Syracuse itself.[61] Neither exposed to the *vulgus*, like the first sphere, nor, like the second, hidden away in an aristocratic house and brought out only for members of the elite with personal connections, Scipio's vision occupies a middle ground. It is open to anyone who reads the *De republica*. Thus it participates in the expansion of the elite that Thomas Habinek sees as so crucial to the transmission of Roman civic values.[62]

Yet this addition of value and expansion through transmission entails risking the loss of control.[63] The two linked visions of the cosmos, Archimedes' and Scipio's, were left behind by men who died violent deaths that no one talks about directly.[64] We have seen that in the *Tusculan Disputations*, Cicero approached the death of Archimedes but replaced it with the image of himself discovering the tomb. In the *De republica*, Cicero alludes to the circumstances of Archimedes' death by introducing the spheres as booty from Syracuse but gives no details about what happened there. Likewise, he refers to Scipio's (future) death but embeds the reference deeply, within a prophecy, which is, in turn, embedded within a dream (6.12). Scipio himself silences his audience's response to the prophecy and refuses to talk about it.[65] Instead, he sets out his vision.

The most valuable prize that the older generation can set out for the younger turns out to be discourse: the dialogue on the state which the narrating Cicero says was set forth to him and Quintus (*exposita*) at Smyrna by Rutilius Rufus; the *ratio* of the sphere which Philus says was set forth to him (*exponere*) in Marcellus's house by Gallus; even the sublime vision of the cosmos, with a promise of a permanent place in the heavens, which Scipio says was set before him in his dream by his grandfather—who told him to commit it to memory (*trade memoriae*)—and which he hands on to his listeners before his death: "When he had spoken thus, I said: 'If indeed a path to heaven, as it were, is open to those who have served their country well, henceforth I will redouble

my efforts, spurred on by so splendid a reward [*tanto praemio exposito*]'"
(6.26). Yet, standing in for stories of death, both the sphere and the vision in
the dream remind us that their creators are dead, just as writing reminds us of
the absence of its author. Neither Archimedes nor Scipio (nor, for that matter,
Cicero) has control over the future of his vision, which after his death depends
on the interest and understanding of others.[66]

Like the metaphors that Gallagher notes, the verb *exponere*, which expresses both the literal ideal of setting something out and the metaphorical
idea of explaining, ties together Archimedes' sphere and Scipio's vision. It is
also used of exposing a child.[67] Scipio says of Romulus:

> is igitur ut natus sit, cum Remo fratre dicitur ab Amulio rege Albano
> ob labefactandi regni timorem ad Tiberim exponi iussus est; quo in
> loco cum esset silvestris beluae sustentatus uberibus, pastoresque eum
> sustulissent et in agresti cultu laboreque aluissent . . . (2.4)

> [*As soon as he was born, then, he is said to have been ordered by the
> Alban king Amulius to be exposed at the bank of the Tiber with his
> brother Remus, because of [Amulius's] fear of his rule being undermined.
> And when he had been nourished in this place at the teats of a wild beast,
> and shepherds had taken him up and reared him in the practices and
> work of rural life . . .]*

The Roman foundation myth, which Scipio presents explicitly as a story useful for the purposes of the dialogue, echoes some of the vocabulary describing the lives of the spheres (*exponi/sustulissent*).[68] It pivots on an idea that ties
together the explanation of the spheres, the foundation story of Rome, Scipio's vision of the cosmos, and perhaps even the publication of the *De republica*. (The verb *exponere* is also used of releasing a book.)[69] Amulius tried to
control the future by exposing the twins, but his loss of control—by the intervention of the gods or fate (Scipio does not say which)—is as crucial a part
of the Roman foundation myth as is the murder of Remus (another death left
unmentioned here). In each case, the person setting out an item loses control
of its future. He cannot control how the knowledge will be used, who will
find the child, who will win the prize, who will read the book. Yet it is this loss
of control, sometimes figured as fate or divine intervention, that makes innovation possible.[70] Thales could not know that Eudoxus would ornament his
invention, that Aratus would transform it into poetry, or how Archimedes
would first replicate and then reinvent it. Socrates could not know what Plato
would do with his ideas or how Cicero, under the guise of replicating them

by "translating" them, would reinvent them as Latin and Roman.[71] Archimedes could not know how Marcellus or Gallus or Cicero would use his sphere. The motif of intervention in the preliminary discussion reinforces this idea. As we have seen, each participant who joined the dialogue both interrupted and continued the conversation, even as he changed its course or expanded its relevance.[72] The marked interruptions suggest that breaks in continuity, whether that continuity is conversational or cultural, can be beneficial—that they bring opportunity, even if that opportunity is only visible from a distant and sometimes alien perspective (perhaps from as far away as the Milky Way). As the *De republica* represents it, the interaction between Greece and Rome is a slowly advancing dialogue, and now it is Rome's turn to talk.

Multiple voices and interruptions are intrinsic to philosophical dialogue, as is narrative embedding.[73] Emblematic of the written philosophical dialogue, Archimedes' two spheres refer beyond the dialogue itself to Cicero's project of creating a Latin philosophical language and to the impact of that project on the Roman Republic and its literature. Looking at the two spheres and the narrative that frames them, we can see Cicero engaged in his own act of Archimedean *inventio*, that of generating cultural capital for himself and for Rome.[74]

Like Vitruvius, Cicero uses Archimedes as a paradigm for discovery. Where Vitruvius rendered Archimedes vividly present at the moment of discovery, by drawing attention to his nakedness and directly quoting his words, Cicero is less direct. Instead of offering an image of Archimedes himself, he presents vivid images of Romans discovering, displaying, and manipulating artifacts associated with him: a Roman seeks out Archimedes' tomb, discovers it, and shares his discovery with the locals; a Roman brings out Archimedes' sphere and sets it in motion. The spheres (by their status as spoils of war) and the grave marker (by its very nature) draw attention to the absence of Archimedes himself. In the case of the tomb, this absence brings up the idea of the heir's responsibilities; in the case of the spheres, it brings up the idea of the role played by time in adding value to and evaluating a prestige item. The description of the first sphere represents an item that has become valuable because of its provenance, whereas that of the second shows an item becoming increasingly prestigious through its associations with important Romans over time. In addition to contributing a touch of realism to Philus's act of recollection— discussing two suns leads to recollecting two spheres—the juxtaposed spheres illustrate a shift in assessing the value of the cultural artifacts, material and textual, taken from a conquered people. The image of the two spheres is emblematic of Cicero's way of casting the Roman appropriation of Greek cultural capital as both inheritance and rediscovery. Thus the two spheres reinforce on many levels the impression of "double vision" that pervades the programmatic passages of the dialogue: two suns, two spheres, two sons (Romulus and Remus), two fathers (Aemilius and Scipio), two ways of appropriating the past.[75]

Finally, in the *Tusculan Disputations,* Cicero calls Archimedes "lowly" (*humilis*), but in the *De republica,* he shows that Marcellus prized his intellectual achievements. Thus Archimedes offers Cicero a paradigm by which even a socially obscure man can find a place in a noble house. Archimedes was important to Cicero partly because he was important to Marcellus. How he was important to Marcellus is a question to which I will turn in part 2.

CODA TO PART ONE
The Afterlife of the Spheres from the De republica

TO SEE EVEN MORE CLEARLY how Archimedes' two spheres together form a figure for the transmission of knowledge, we can examine the discussion of a sphere that appears in a work from late antiquity, the *Matheseos libri VIII*, eight books of astrological learning written by the educated Sicilian aristocrat Julius Firmicus Maternus.[1] This treatise, also called the *Mathesis*, includes a horoscope of Archimedes, to which I will refer elsewhere;[2] my argument here draws on the material that frames the astrological information, that is, the prefatory and concluding remarks and the occasional asides that Firmicus directs toward his dedicatee. Although Firmicus's treatise is not a dialogue, these passages recall the framing material of the *De republica*; and the preface, which, in Jean Rhys Bram's words, "bristles with literary allusions," reworks Cicero's sphere passage to its own ends.[3]

Firmicus presents the *Mathesis* as the fulfillment of a promise made some time earlier, when, after a long and debilitating journey, he was nursed back to health by one Lollianus Mavortius, a government official in charge of Campania.[4] While he was convalescing, says Firmicus, he and Mavortius fell into conversation about their lives and deeds and exchanged "stories of conversations" (*sermonum fabulas*).[5] Mavortius asked Firmicus about his native land, the topography of Sicily, and topics long famous, ranging from a rationalization of the myth of Scylla and Charybdis to explanations of natural phenomena seen at Aetna and Lake Palicus.[6] Thus, like the early passages of the *De republica,* the *Mathesis* invokes the memory of a conversation that took place long before, and it, too, involves further layers of memory: as both men recalled the past (pr. 4), "recollecting in turn and recalling to memory long-ago deeds" (*recolentes invicem pristinos actus et ad memoriam revocantes*), Mavortius asked Firmicus about "everything that the old stories transmitted" (*et omnia quae veteres fabulae prodiderunt*). Mavortius, says Firmicus, even remembered traditions passed down by the writers he read as a child.

> Recollecting [*recolens*] with me and looking into everything else that Attic and Roman literature [*litterae*] handed down [*tradiderunt*] to you from childhood about the wonders of the province of Sicily, to crown all, you shifted [*transtulisti*] the thread of your conversation and speech to the sphere of Archimedes, displaying [*ostendens*] to me, as you did so, the wisdom [*prudentiam*] and learning [*doctrinam*] of your divine genius [*divini ingenii*]. (5)

According to Firmicus, Mavortius's quest for knowledge does not come from nowhere; it has its source in old stories or childhood reading, that is, in the memory of prior knowledge.

Within this account of remembered acts of remembering, a pair of transitions takes place. Mavortius shifts the conversation to Archimedes' sphere and assumes the role of informant, as he explains or identifies the uses of the nine spheres (*globi*) and their five zones; the twelve signs of the zodiac; the effects of the five planets; the daily and annual paths of the sun; the movement and waxing and waning of the moon; the revolutions needed to make the greater year; the nature of the Milky Way and of solar and lunar eclipses; why the Dipper never sets; what part of the earth receives the north wind, what part the south; how the earth is suspended in the very center; and why Ocean surrounds the land as if it were an island. Mavortius, then, explains the sphere to Firmicus, as Gallus did to Philus in the *De republica,* with, of course, some significant differences.

On the one hand, the long series of specific identifications and explanations produces a description of a sphere far more detailed than the one in the *De republica*; on the other, the sphere is not represented as being present when Mavortius explains it, as it was in the *De republica.* Yet just as the second sphere was brought forth for display in the *De republica,* so, too, in this passage something is brought forth for display, not the artifact that shows Archimedes' talent (*ingenium*) to be beyond human, but, rather, Mavortius' wisdom, learning, and divine talent (*divini ingenii*).[7] It is Mavortius's wisdom and learning, displayed through explanation, that generates the dialogue.

> Since all these were being passed on to me [*traderentur*] by you, Mavortius, ornament of good men, by the deft instruction of your demonstration [*facili demonstrationis magisterio*], I myself even dared to offer up [*proferre*] something with the rashness of ill-considered speech, to promise [*promitterem*] to produce for you whatever the wise Egyptians of old, and the divine men, the sages of Babylon, passed on [*tradiderunt*] to us by the instruction [*magisterio*] of their divine learning, about the force and powers of the stars. (1 pr. 6)

Firmicus has read his Cicero closely. His representation agrees that the discourse on the spheres is of greater importance to the dialogue than are the spheres themselves. Mavortius's display elicits the promise of a treatise, because it proves him to be a suitable recipient of astrological learning. As Tamsyn Barton points out, the dominant idea in the framing passages is the handing over of arcane knowledge in the manner of an initiation.[8] Having a suitable audience is the necessary condition for transmitting arcane knowledge. Just as Gallus ordered the sphere emblematic of arcane knowledge to be brought forth (*iussisse proferri*, Cic. *Rep.* 1.21), so, too, Firmicus says that he has ventured to bring forth the promise of his arcane knowledge (*ausus sum. . . . aliquid . . . proferre*), now that he has found Mavortius to be both eager for information and, what is more important, skillful at passing on what he has learned. (The phrase *facili demonstrationis magisterio*, then, does more than simply flatter Mavortius.)[9]

The author cringes a great deal in this preface—and indeed in all the framing material—as he speaks of the uncertainty born of his modesty, his nervousness, and the frailty of his talent (1 pr. 1). But as he says repeatedly (1 pr. 1, 6–8; see also 8.33.1), he has made a promise, which he feels compelled to keep, especially in response to Mavortius's urging and encouragement.[10] That encouragement and that urging arise from what Firmicus presents as yet another act of transmission.

> For when the clear and admirable judgments of our master and emperor Constantinus Augustus had handed over to you [*tradidissent*] the steering oars [*gubernacula*] of the entire West, you allowed no time to pass in which you did not demand from us what we had rashly promised. (1 pr. 7)

Thus several acts of transmission have together brought about the present situation: Mavortius has handed over an interpretation of the sphere to Firmicus; Firmicus has reciprocated by promising to hand on to Mavortius the knowledge of astrology, which the Egyptians and Babylonians have handed to him; Constantine has handed over the *gubernacula* to Mavortius, who responds by demanding that Firmicus fulfill his promise and hand over the promised knowledge. The many uses of *tradere* in the various framing passages at the beginnings and endings of books testify to Firmicus's interest in the transmission of arcane knowledge.[11] The use of *transferre* to mark shifts in topic throughout the treatise recalls the use of *transferre* to describe Mavortius's shift of the topic to Archimedes' sphere.[12] In addition, these verbs connect the sphere to the ideas of shifting discourse and the transfer of arcane knowledge that are conjoined in Firmicus's talk of the gradual revelation of information to Mavortius as to an initiate. The many references to promising (*promittere*) also help to connect the preface to the other framing passages in the treatise.[13] But the result of the promise and the means of transmission will be through books (*libelli*, 1), which will reach a wider audience than the initiate alone, an audience that can include even the emperor, the recipient of panegyric at the end of book 1. By thus pointing to the expansion of even arcane knowledge, Firmicus's preface connects his act of transmission to the one at the heart of the *De republica*.

There were, of course, other obvious reasons for Firmicus to open his work with the image of Archimedes' sphere: it reflects his patriotism as a Sicilian, and by invoking the image of the Ciceronian spheres, which itself anticipates Scipio's Dream, Firmicus doubly anticipates the privileging of the planets in his astrological system.[14] But using the image of the sphere in a context that so emphasizes transmission gives that image all the more meaning, for it suggests that the *Mathesis* intends to transmit a complete vision of the cosmos as does the *De republica*, and it restates Sicily's claim to be at the center of learning—learning redefined, by Cicero, as the reception and transmission of knowledge rather than individual, isolated, discovery.[15] The Sicilian native, then, presents the conversation between himself and the Roman official Mavortius as dialogue, which performs the same role in producing cultural capital as did Cicero's description of two spheres.

PART TWO

A SKETCH OF EVENTS
AT SYRACUSE

I INCLUDE HERE, simply to orient the reader, a brief sketch of events at Syracuse, based largely on Polybius, Livy, Cicero, and Plutarch, some of the very sources discussed in the following chapters.[1]

Hieron II died in 215 BCE, having ruled Syracuse for over fifty years. During his long reign and at his urging, Archimedes developed the city's defense engines and fortifications. Hieron had made an alliance with the Romans during the first of the Punic wars and proved himself loyal during the early years of the second, supplying grain, money, and troops at critical points. Because his son, Gelon, predeceased him, the rule of Syracuse came to his grandson, Hieronymus, together with a group of advisors.

After the disastrous Roman defeat at Cannae, Hieronymus made an alliance with Carthage. He did away with the advisors who advocated loyalty to Rome, and if Livy is to be believed, he displayed all the stereotypical traits of an evil tyrant. This provoked a conspiracy against him, and he was assassinated. When the pro-Carthaginian faction attempted a coup, the city massacred the entire royal family and took some tentative steps toward constructing a stable government and renewing its ties to Rome.

By late 214, one of the Roman consuls, M. Claudius Marcellus, had been appointed to his command in Sicily and was engaged in negotiations with Syracuse. In response to (perhaps) exaggerated accounts of atrocities that he had perpetrated elsewhere on the island, the pro-Carthaginian element in Syracuse seized power again, and the city closed its gates to the Romans. In the spring of 213, Marcellus and the praetor Ap. Claudius Pulcher attacked Syracuse by land and sea. Their failure to take Syracuse with their first, mighty assault is credited to the defense engines of Archimedes. Having given up the assault, the Romans began to besiege the city. They captured part of it by a night attack during a religious festival and took the rest only late in 212, when its Carthaginian defenders were severely afflicted by a plague.

When Syracuse finally was captured, Marcellus, so the story goes, stood on the walls and wept, thinking of the city's illustrious past and the beauty that he was about to destroy. He decided to save the city's physical fabric from destruction by fire. When the people of Syracuse sent envoys asking for mercy, Marcellus ordered either, generally, that no free citizen be harmed or, specifically, that Archimedes be spared; he then turned the city over to his soldiers to plunder. In the chaos that accompanied the sack of Syracuse, however, a soldier who came upon Archimedes (who was intent on his diagrams) did not know who he was and killed him. Marcellus grieved at this and saw to Archimedes' burial. He also saw to the transfer of Syracuse's portable works of art to Rome. Among this loot were, possibly, Archimedes' spheres, one of which ended up in the temple of Virtus, the other in Marcellus's house.

CHAPTER 4

WHO KILLED ARCHIMEDES?

Valerius Maximus, author of the nine-book *Memorable Deeds and Sayings* (*Factorem et dictorum memorabilium libri IX*), tells the story of Archimedes' death as an example of extraordinary *industria*, or "diligence."[1]

Archimedis quoque fructuosam industriam fuisse dicerem, nisi eadem illi et dedisset uitam et abstulisset: captis enim Syracusis Marcellus machinationibus eius multum ac diu uictoriam suam inhibitam senserat, eximia tamen hominis prudentia delectatus ut capiti illius parceretur edixit, paene tantum gloriae in Archimede seruato quantum in oppressis Syracusis reponens. at is, dum animo et oculis in terra defixis formas describit, militi, qui praedandi gratia domum inruperat strictoque super caput gladio quisnam esset interrogabat, propter nimiam cupiditatem inuestigandi quod requirebat nomen suum indicare non potuit, sed protecto manibus puluere 'noli' inquit, 'obsecro, istum disturbare', ac perinde quasi neglegens imperi uictoris obtruncatus sanguine suo artis suae liniamenta confudit. quo accidit ut propter idem studium modo donaretur uita modo spoliaretur

I should say that Archimedes' diligence too was fruitful, had it not both given him his life and taken it away: for at the capture of Syracuse, Marcellus had perceived that his own victory had been greatly and long delayed by Archimedes' machines; still, pleased by the man's outstanding proficiency, he decreed that Archimedes' life be spared, placing almost as

much glory in saving Archimedes as in crushing Syracuse. But Archi-
medes, while with mind and eyes pegged to the earth he drew diagrams,
because of his excessive interest in pursuing his objective, could not tell his
name to a soldier who had broken into his house to pillage and who, with
sword drawn over his head, was asking him who he was. Rather, protect-
ing the dust with his hands, he said, "I beg you, do not disturb it." And,
slaughtered as disregarding the victor's command, he disordered the out-
lines of his own art with his own blood. Thus it came about that on ac-
count of the same zeal he was now granted life, now despoiled of it.[2]

Vivid, memorable, heavy with moralizing irony, this image of the gallant old
mathematician dying in defense of geometry became the canonical version
of the story in the Middle Ages, inspired Petrarch (see chap. 6), and captured
the imagination of artists in the Renaissance and early modern era.[3]

We can see some of Valerius's famous verbal agility at work in this carefully
constructed passage. In addition to the smooth transition from the previous
exemplum, there is also ring composition: by opening and closing the anecdote with references to the lethal effects of *industria,* Valerius emphasizes his
point and sets this example apart from the others.[4] As well he might, for he
presents *industria* positively in all his other examples of it, seven Roman and
sixteen foreign;[5] he introduces only Archimedes' *industria* negatively, as a
quality that he would call profitable had it not killed its possessor.[6] In addition, the anecdote elides the actual act of killing, as the main verbs move from
Archimedes' penultimate deed, an ineffective speech act ("'Do not,' he said, 'I
beg you, disturb it'" [*'noli,' inquit 'obsecro, istum disturbare'*]), to his ultimate,
his disordering the lines of his own diagrams with his blood (*confudit*). The
reference to the actual killing appears between the two actions, in the passive
participle *obtruncatus* ("slaughtered") with no agent. Valerius leaves it to readers to make the connection between the raised sword, the participle, and the
spilled blood.[7] These dots are not difficult to connect, and the passive voice of
the participle keeps Archimedes in the foreground; still, the syntax leaves a gap
that, in this context, is meaningful—enough, at least, to leave an experienced
reader of rhetorical school texts with a reasonable doubt as to the soldier's
guilt.[8] Not only does the anecdote elide an important event in the story—the
very event, in fact, that gave rise to the story—but in the introduction and
conclusion that frame the narrative, Valerius also firmly states who is responsible for this death: no soldier bent on pillage, but that violent and dangerous
abstraction Industria and her accomplice Studium.[9]

Finally, and most important for the argument that this chapter makes,

Valerius tells the story without mentioning any sources. Speaking in his own person, he is both its narrator and the authority responsible for it. He even quotes Archimedes directly, albeit in Latin, not the Doric Greek of Syracuse. Valerius, in fact, comes as close as—perhaps closer than—any other surviving ancient author to taking responsibility qua narrator for the account of Archimedes' death. Yet at the same time, this version explicitly places the moral responsibility for the death itself on Archimedes' own character.[10]

Arguing that this inverse relationship of narrative to moral responsibility appears consistently in the sources is my preliminary task in this chapter, which has three principle and related aims. The first is to disclose the strategies by which several ancient writers dealt with the same apparently regrettable incident, or, to be more precise, to disclose the strategies by which several ancient writers represented the same incident as regrettable. Another is to explore how the various contexts in which this incident is embedded give a different shape to each version of the story. The third is to show how a narrative that invokes tradition by flagging its own status as an exemplum and mobilizing the dynamics of intertextuality uses that tradition in ways that serve or criticize the ideology of empire. Although, to paraphrase Andrew Laird (Laird, *Expressions of Power*, xiii), there is no "control" version of this story from which the others deviate—its sources are distributed across several centuries and genres, and different writers use the exemplum for different purposes and in different contexts—still, when we review the various versions, we see can one narrative gesture appear repeatedly. It is a movement of evasion, and it takes different forms depending on the rhetorical situation: either authors make someone else responsible for the story, as if to say, "I am not the messenger—he is"; or, like Valerius, they elide the act of killing; or they avoid referring to it altogether; or, again like Valerius, they shift the blame for Archimedes' death onto Archimedes himself.

Writing a hundred and forty-two years after the event, Cicero provides our earliest surviving reference to Archimedes' death. The event appears in a passage from the fourth part of the second *actio* against Gaius Verres, the famous speech about the art treasures that Verres stole from Sicily when governor there.[11] The passage's immediate goal is to emphasize Verres' greed at Syracuse by conjuring up in contrast the memory of the restraint shown by Marcellus after he captured the city during the Second Punic War. Syracuse is the climactic example in a long litany of pillage, and in the section that relates its plundering, Cicero commands (*conferte*) the jury to juxtapose Verres' behavior as governor to Marcellus's as conqueror (*Verr.* 2.4.115–31). Cicero introduces the topic by bringing up Marcellus as the contrasting exemplum; then he gives an overview of Verres' atrocities toward the Syracusans via a *praeteritio,* describes Syracuse, and recounts the theft of particular items, concluding with a statue of Jupiter; he rounds off his discussion of the sack by returning to Marcellus and telling of Archimedes' death.[12] Archimedes' death, then, is the climactic exemplum in this subsection of the larger argument.

> Ut saepius ad Marcellum revertar, iudices, sic habetote, pluris esse a Syracusanis istius adventu deos quam victoria Marcelli homines desideratos. Etenim ille requisisse etiam dicitur Archimedem illum, summo ingenio hominem ac disciplina, quem cum audisset interfectum permoleste tulisse: iste omnia quae requisivit, non ut conservaret verum ut asportaret requisivit. (*Verr.* 2.4.131)

> *[Although I return to Marcellus rather often, gentlemen of the jury, consider this: that the Syracusans lost more gods by Verres' arrival than they did men at Marcellus's victory. For he, in fact, is said even to have sought out the famous Archimedes, a man of the highest talent and learning, and when he heard that he had been killed, to have born it very ill. But Verres sought all that he sought not in order to save it but in order to carry it away.]*

Of immediate interest to us here is that Cicero reports neither the killing itself nor even the report of the killing. He reports the report of Marcellus's reaction to the report. "He ... is said" (*ille ... dicitur*) is the main clause of the entire sentence. As in Valerius, here, too, the passive (*interfectum*) makes it unnecessary to identify a killer. Of course, when one is emphasizing Marcellus's clemency and restraint, one does not want to depict vividly a murder scene. I will deal later with the effect this passage produces by bringing up the death of Archimedes at all.

In a passage from the *De finibus* that may have inspired Valerius, Cicero has been discussing the compelling nature of the love of learning. After pointing out how Odysseus listened to the Sirens, he concludes:

> It is knowledge they [the Sirens] promise, and it was no wonder that it was dearer than his fatherland to a man eager for wisdom. And yet, to desire to know all things, of whatever sort they be, is the mark of the inquisitive, but indeed to be led by the thought of greater things to the desire of knowledge must be considered the trait of the highest men. (5.48–49)[13]

Cicero then lists examples of these *summi viri,* the first of whom is Archimedes (5.50): *Quem enim ardorem studii censetis fuisse in Archimede, qui dum in pulvere quaedam describit attentius, ne <quidem> patriam captam esse senserit?* (For what burning desire for study do you think had been in Archimedes, who, while he wrote something in the dust quite carefully [or "too carefully"], did not even perceive that his fatherland had been captured?). But instead of going on to report Archimedes' death, Cicero turns to the next exemplum, that of the hardworking Aristophanes. The repeated reference to the fatherland and the suggestion that Archimedes studied his figures with, perhaps, excessive concentration places the mathematician in the same category as those who, by succumbing to the Sirens, showed that they found knowledge more alluring than their *patria.* Thus Cicero gives the impression, albeit faint, that while such burning desire for knowledge is characteristic of the greatest men, it also might be considered unpatriotic.

Like Cicero's, Livy's account of the death presents not the event but the report of the event.

. Cum multa irae, multa auaritiae foeda exempla ederentur, Archimedem memoriae proditum est in tanto tumultu, quantum captae urbis in <uiis> discursus diripientium militum ciere poterat, intentum formis quas in puluere descripserat ab ignaro milite quis esset interfectum. (25.31.9)

[While many foul instances of anger, many of greed were being carried out, it is handed down to memory that Archimedes, in such chaos as the running around of soldiers engaged in plunder in the streets of a captured city could stir up, focused as he was on the shapes that he had drawn in the dust, was killed by a soldier who did not know who he was.]

Once again the passive voice keeps the figure of Archimedes to the fore (*Archimedem . . . intentum . . . interfectum*). But now a killer has surfaced. We do not know who he was, and his one distinguishing feature is that he did not know who Archimedes was. Does this make him more or less responsible? Livy's soldier did not ask the identity of his victim. In Valerius, the soldier at least asked.

Running through a list of learned men in book 7 of his *Naturalis historia,* Pliny the Elder makes only a rather sinister and almost untranslatable statement concerning Archimedes: "A weighty proof (*testimonium*) of Archimedes' knowledge of geometry and mechanics befell him by the decree of M. Marcellus, at the capture of Syracuse, that he alone not be injured, had not the thoughtlessness of a soldier proved the command to be in vain."[14] Like Valerius, Pliny "fingers" a killer abstraction. But this time, the perpetrator of the murder is not Archimedes' *industria* but a soldier's "thoughtlessness" (*inprudentia*). Silius Italicus elides the death completely, as does Firmicus Maternus. (I will discuss the passage from Silius Italicus later in this chapter.) Firmicus Maternus says of Archimedes—in Archimedes' horoscope—that "[he] often laid low Roman armies with his mechanical skills"; he then adds, "Marcellus, during the triumph for his victory, amongst the exulting cries of his soldiers and the triumphal laurels, mourned this man with grievous sorrow."[15] Readers can be excused for wondering if they missed a step. Why did Marcellus mourn this man? What killed him—old age, plague, an accident with a catapult?

Plutarch at first appears to be the exception to this inverse relationship of responsibility for deed and narrative. He both blames a Roman and presents Archimedes' death straightforwardly.

> For he happened to be by himself, working out some problem with the aid of a diagram [ἔτυχε μὲν γὰρ αὐτός τι καθ᾽ ἑαυτὸν ἀνασκοπῶν ἐπὶ διαγράμματος], and having fixed his thoughts and his eyes as well upon the matter of his study, he was not aware of the incursion of the Romans or of the capture of the city. Suddenly a soldier came upon him and ordered him to go with him to Marcellus. Archimedes refused to do this until he had worked out his problem and established his demonstration, whereupon the soldier, angered [ὀργισθείς], drew his sword and dispatched [ἀνεῖλεν] him.[16]

We now have a weapon, a motive, even a soldier who is the subject of an active verb in a direct statement. Finally, we seem to have found an author who is firm about who killed Archimedes. But Plutarch continues, ἕτεροι μὲν οὖν λέγουσιν (others, however, say)

> that the Roman came upon him with drawn sword threatening to kill him at once [ὡς ἀποκτενοῦντα] and that Archimedes, when he saw him, earnestly asked him to wait a little, that he might not leave the result that he was seeking incomplete and without demonstration; but the soldier ignored his request and finished him off.

Once again the Roman soldier plays an active role. In fact, in this version, he approaches Archimedes with intent to kill. Yet Plutarch both presents this version as an alternate to the first and embeds it within an indirect statement; moreover, he continues, "and there is yet a third story" (καὶ τρίτος ἐστὶ λόγος):

> that as he was carrying to Marcellus some of his mathematical instruments, such as sundials and spheres and quadrants, . . . some soldiers fell in with him and, thinking that he was carrying gold in the box, slew him.

Plutarch's three versions form a progression, starting with a soldier who did not intend to kill Archimedes at first (what might be categorized in modern terms as "unpremeditated homicide"), advancing to a single soldier who intended to kill Archimedes from the start ("premeditated homicide"), and culminating in a group of greedy soldiers shamelessly attacking one old man (homicide and robbery). The behavior of Roman soldiery seems to go from bad to worse. But at the same time, the three stories form a regression, as it were, in terms of narrative embedding, from a narrative presented in the author's voice; to that of ἕτεροι and their alternate voices, presented in indirect speech; to the death introduced explicitly as story (a λόγος), also presented in indirect speech. As in Valerius's account, here, too, we can see an inverse relationship between narrative and moral responsibility: the less Plutarch takes responsibility for the narrative, the more moral responsibility a Roman (or group of Romans) takes for the act. At the end, by calling the last version a τρίτος λόγος, Plutarch reveals all three as λόγοι, no one of which can claim to represent how Archimedes' death "really happened."

Why do we find all this narrative hedging? Why does no one represent the death both completely and directly? The sources agree on the central point, that Archimedes died during the Roman sack of the city; and as we shall see in the next chapter, they are not nearly so reticent about either Archimedes' role in the defense of Syracuse or his demoralizing effect on the Roman attackers. Polybius, Livy, Plutarch, Valerius, and Pliny all agree that Archimedes' efforts, more than anything else, delayed the capture of the city. The most detailed versions of the defense—those of Polybius, Livy, and Plutarch—report his actions in a relatively straightforward way.[17] It seems inconsistent that the sources directly present Rome's inability to overcome Greek machines with military force yet shy away from representing explicitly the death of the enemy who made those machines, especially since Rome was fighting for its existence in the Second Punic War and the killing was an act of war, not a peacetime murder.

This reticence may, of course, be a mere accident of survival. Perhaps a lost passage from Polybius's history included a detailed death scene that assigned a Roman blame.[18] Still, the fact remains that, despite numerous opportunities to report Archimedes' death in a straightforward way, one that would take full authorial responsibility for the story and make a Roman initiate the killing, no surviving ancient source does so. Zonaras (perhaps preserving a detail from Dio) presents a defiant Archimedes: "'Let them attack my head,' he said, 'but not my drawing.'"[19] Tzetzes (perhaps preserving a detail from Dio or Diodorus) even makes Archimedes draw his weapon first: "Give me one of my machines!" cries Archimedes, when he perceives that a Roman is upon him.[20] From a modern point of view, it almost appears as if these writers are defending Rome against a charge of premeditated homicide on various grounds: no one really knows what happened; the Romans didn't *mean* to kill him; they didn't mean to kill *him*; it was a matter of self-defense; the victim was to blame.[21]

An interpretation that comes immediately to mind is that these omissions and embeddings are evasive strategies that help the authors cope with an event that became a focal point of Roman shame and regret. According to this line of interpretation, the death story took shape as what Matthew Roller calls a "fig leaf," concealing, in this case, not a Roman defeat but the shameful violence that accompanies conquest.[22] This is perhaps part of the answer. Killing an old man who is not even paying attention is nothing of which to be proud. Consider how the sources represent the killer: Cicero leaves him out; Livy and Valerius say that he did not know who Archimedes was. (He apparently lacked the education in geometry that would have allowed him to recognize the mathematician by his very act of drawing in the dust.)[23] Pliny blames the soldier's thoughtlessness; Plutarch says that Marcellus later shunned him as if he were polluted (ἐναγή); the twelfth-century Tzetzes goes so far as to say, "As for the killer, Marcellus had him beheaded, I think."[24] No one preserves his name.[25]

But discomfort arising from shame need not be the only explanation. Cicero's version says explicitly that the fall of Syracuse was a story known to his audience, and he implies that the anecdote about the death (from here on, I will call it the "Death of Archimedes") is a familiar exemplum.[26] As an exemplum, it can be manipulated to serve various purposes, one in the *Verrines,* another in *De finibus.* Accordingly, we need to consider it as part of what Roller has called "the discourse of exemplarity," a discourse that exploits the simultaneous familiarity and flexibility of the past. Cicero, Livy, and Plutarch either explicitly or implicitly note the story's status as part of a preexisting tradition, and this exemplary status can also help to explain such pregnant omissions as those of Valerius, Pliny, and Firmicus Maternus. The "Death of Archimedes" is simply too familiar to need retelling in full. The question arises, then, of whether the sources reflect regret about a shameful event or represent regret. A view of the story as a "fig leaf" takes them as reflecting regret. If we take them as representing it, we need to ask another legal question: *Cui bono?*[27] To whose advantage is not the death of Archimedes but the "Death of Archimedes," not the event but the exemplum? To whose advantage is the representation of regret?

So far, I have focused on victim, killer, and narrator. But we have not far to look to see that another party is involved. Most sources for the death refer in some way to the Roman conqueror of Syracuse, Marcus Claudius Marcellus: Cicero, as we have seen, presents the story as one of Marcellus learning the bad news; Livy, too, after telling of Archimedes' death, reports Marcellus's reaction. Livy says—still as part of what was "handed down to memory" (25.31.10): *aegre id Marcellum tulisse sepulturaeque curam habitam, et propinquis etiam inquisitis honori praesidioque nomen ac memoriam eius fuisse* (Marcellus bore this ill and saw to his burial, and for Archimedes' connections, moreover, after they were sought out, his [Archimedes'] name and memory were a source of honor and protection). In Cicero and Livy, Archimedes' death is subordinate to Marcellus's story. Pliny, too, presents Archimedes' death as an event in the life of Marcellus, the unintended voiding of Marcellus's decree that Archimedes be spared. Valerius Maximus says that Marcellus, pleased by Archimedes' exceptional intelligence, ordered that he be saved.

In like manner, Plutarch introduces his three accounts with the following words: "but what afflicted *Marcellus* particularly was the death of Archimedes" (*Marc.* 19.4, emphasis added). Then, after running though all three λόγοι, he concludes with the one undisputable point: "however, what is agreed [ὁμολογεῖται] is that Marcellus was afflicted [ἤλγησε] at Archimedes' death, and turned away from the killer as from one polluted, and, having sought out Archimedes' kin, paid them honor" (19.6). Moreover, Plutarch gives us the closest text we have to a surviving ancient biography of Archimedes, a pair of digressions from the life story of Marcellus, and these digressions support Plutarch's assertion, at the beginning of the biography, that Marcellus loved Greek culture (see chap. 5). Finally, both Silius Italicus and Firmicus Maternus refer to Archimedes' death only by way of Marcellus's grief. In short, the tradition not only displays a certain reluctance to represent Archimedes' death explicitly; it also presents that death as embedded deeply in the story of Marcellus: Archimedes' death caused Marcellus pain, made Marcellus cry, violated Marcellus's decree, and marred his moment of triumph. That anonymous Roman soldier, then, plays an important narrative role, for he both keeps Marcellus from being directly responsible for the killing and brings it about that the name most closely associated with it is that of Marcellus.

Read with an eye to its embeddedness and omissions, the "Death of Archimedes" becomes an even more interesting narrative, one that is revealed as "angled, selective, partial and ideological."[28] The most striking result of eliding or de-emphasizing the death is not to obscure Roman responsibility but to place emphasis on Marcellus's reaction. It is essentially a story about the reception of the story, reception that indicates the measure of both Archimedes' extraordinary genius, by reporting his enemy's high esteem of him, and the taste, education, and intelligence of the enemy who made such an evaluation.[29] As a story about Archimedes, then, it is a fine story about Marcus Claudius Marcellus, five times consul, victor many times over in single combat (most famously at Clastidium), dedicator of the *spolia opima*, conqueror of Syracuse, and member of a Roman family that was prominent in politics and literature in the third and second centuries BCE and in Augustus's time.[30] In fact, in most elements of the anecdotal and exemplary tradition—the stories of sphere, death, defense of Syracuse, even the tomb—Archimedes is important because of his indirect relationship with this man, who has been called "the quintessential Roman" and whose legacy was hotly contested in the century after his death.[31]

That the avowed philhellene Marcellus was responsible, even only indirectly, for the death of the Greek genius was irony of the kind that his political enemies would have relished and could have exploited.[32] They might have pointed out that although Marcellus brought all that pernicious art to Rome, he failed to bring Archimedes. Polybius, who is hostile to Marcellus elsewhere, may have represented Archimedes' death in a manner critical of Marcellus;[33] he does criticize the impractical nature of the spoils that Marcellus brought to Rome, on the grounds that paintings and sculpture do no good in war (*Polyb.* 9.10). Polybius or even Cato could have pointed out that a live Archimedes, the Werner Von Braun of his time, would have made a greater contribution to Rome's effectiveness against Hannibal than could statues and paintings.

Some traces may even remain of the idea that Archimedes was the "one that got away." Cicero's presentation of the death in the *Verrines* suggests that Archimedes would have been a particularly prestigious war prize. Archimedes does not appear among the humanitarian categories of those wronged at Syracuse (*Verr.* 2.4.116); instead, he appears after Cicero has said that he will return to his main topic, stolen art (2.4.131). While the repetition of the verb *requiro,* "to seek" (*etenim ille requisisse etiam dicitur Archimedem illum,* 2.4.131), helps heighten the contrast between Marcellus's well-meaning, if futile, search for Archimedes and Verres' greedy and productive sweep, it relies on a parallel between Archimedes and art as objects of a Roman search. Marcellus's thwarted purpose in seeking Archimedes is embedded in Verres' achieved mission of carrying away *things: iste omnia quae requisivit,* <u>*non ut*</u> <u>*conservaret*</u> *verum ut asportaret requisivit.*[34] Other faint suggestions that Archimedes would have been a particularly splendid prize may remain in Firmicus Maternus, who writes of Marcellus mourning for Archimedes during his very triumph.[35] Finally, Livy's including siege engines among the prizes displayed in Marcellus's ovation makes the absence of their dead inventor all the more conspicuous.[36]

Although there is no glory in killing Archimedes, with the right spin, there can be glory in mourning him. The existing "Death of Archimedes" bears the marks of a pro-Marcellus tradition. The Sicilians did become clients of the Marcelli, and Cicero, the source of the earliest version of the death, reports it in a passage in which he aligns himself with Marcellus as their *patronus;* all the other versions of the "Death of Archimedes" are Augustan or later and thus date from the time when Julia's marriage to Marcellus had brought the family back into prominence.[37] Although traditions hostile to Marcellus surface in other stories about his life, this story probably survives in this version as a result of a circular process by which, as an expression of benevolence toward a conquered people, it contributed to the tradition that won. Whatever Marcellus's real feelings about Archimedes were— and his grief might well have been sincere—our reading of the "Death of Archimedes" allows us a glimpse of how one of the great noble families of the middle and late Republic may have exerted control over an event that was potentially damaging to its reputation, turning the story to its advantage.

The "Death of Archimedes" preserved by Valerius, Pliny the Elder, and Firmicus Maternus includes the immediate context of the capture of Syracuse but not the larger historical context of the Second Punic War.[38] Cicero's *Verrines,* to some degree, and, of course, Livy and Silius Italicus, to greater degrees, all present the "Death of Archimedes" as part of extended war narratives. I focus on these next, not just because they present the story in the larger context but also because their versions of the story emphasize the role of tradition (for more discussion of Plutarch, see chap. 5). In addition to insisting that Archimedes was killed against Marcellus's wishes, the sources also report that Marcellus did not see the killing but only heard about it. Thus they connect Marcellus to the news of the death rather than to the death itself. This emphasis on Marcellus hearing, not seeing, parallels and underscores the story's many references to tradition and commemoration: Marcellus hears about Archimedes' death and commemorates him with a funeral; writers hear about Marcellus's hearing about the death and commemorate his response by using it as an exemplum; Cicero's audience hears or reads the story, and readers, such as Livy and Silius Italicus, read it and pass it on.

That these stories join a long and distinguished tradition of Syracusan historiography becomes especially emphatic at the city's capture.[39] Cicero introduces the passage that tells of it with an ambiguous statement: "I shall, moreover, gentlemen of the jury, recount the plunder of the single most beautiful and most magnificent city of them all, Syracuse, and make it known, in order finally to bring to a close and set a limit to all speech of this type."[40] The first words identify the topic to be treated as the *direptio* of Syracuse. The phrase "I shall recount the plunder" (*direptionem commemorabo*), together with the superlatives "most beautiful" (*pulcherrimae*) and "most magnificent" (*ornatissimae*), suggest that Cicero will retell the story of Marcellus's conquest, which took place when the city was indeed *pulcherrima* and *ornatissima* because not even Marcellus, to say nothing of Verres, had sacked it and removed its art to Rome. Cicero's next words reinforce this expectation by referring to the audience's familiarity with the old story: "There is almost no one of you but has often heard how Syracuse was captured by M. Marcellus, no one but has at some point read it in the annals."[41] Cicero claims that his audience knows the story from multiple sources, both oral and written. One might expect that a retelling of the story would follow, an explication of how Syracuse was captured by M. Marcellus, but this is only a feint. The *direptio* that follows is Verres' plunder.[42]

This passage's immediate rhetorical mission, to emphasize Verres' rapacity by conjuring up the contrasting memory of Marcellus, is complicated by the need to conjure up the sack of Syracuse without actually retelling the story. The sack, after all, was violent and bloody and ended in a massive transfer of wealth to Rome.[43] Moreover, one of the Marcelli, C. Marcellus, proconsul of Sicily in 79, was among the *iudices* to whom the speech was addressed, and another was lined up to testify for the defense.[44] The story clearly required careful handling. Cicero negotiates these shoals by presenting the capture of the city not as an event but as remembered discourse.[45] Instead of listening to Cicero's narrative of Marcellus's sack, the listeners are to conjure up their own complete and preexistent memories from their experience of the multiple oral and literary sources and set them against the story that Cicero will tell. Moreover, this hypothetical narrative that the audience is to construct from a shifting, amorphous, and polyphonic tradition receives all its shape from Cicero's sharp contrast with the plunder by Verres: "Set this peace against that war, this praetor's arrival against that general's victory, this man's foul band against that man's unconquered army, this man's lust against that man's restraint."[46] The speech instructs the members of the jury to recollect not a particular older narrative but, rather, elements of one (war, victory, an unconquered army, and restraint), which they are to reassemble and reorganize in their thoughts according to Cicero's antitheses.

Cicero presents Marcellus's and Verres' behavior as opposites that will, he says, exchange positions in the minds of his audience. His comparison represents Marcellus's famous sack of Syracuse as the less extreme event of a pair and then, it turns out, not a sack at all. If the jury compares Marcellus and Verres and responds as Cicero expects, it will literally redefine both events: "You will say that Syracuse was founded by him who captured it, but captured by him who received it when it had been established."[47]

Cicero's introduction establishes the sharpest contrast possible between two stories of the sack of Syracuse, between Marcellus and Verres, between what "almost everyone knows" and what Cicero will bring forth. The old story lies in the distant, as opposed to immediate, past. It can be experienced only indirectly, via reading or hearing the historical tradition, whereas Cicero has himself seen what Verres has done and will place it before the jury's eyes. Cicero's comparison compels his audience to adjust its memories so as to meet the points of contrast as he sets them out. Thus, conjuring up this fluid tradition of Marcellus' sack of Syracuse, Cicero produces a rereading of all the narratives of the sack that are available to his audience.

Cicero introduces the old story with an eye to its status as part of the historical tradition and uses this idea in a way that aligns Marcellus and the audience of the speech. *Verrines* 2.4.115–31 closes as it began, with reference to the tradition (*dicitur*), but also with the repetition of *audire*, this time used not of Cicero's listeners but of Marcellus.[48] The appearance of *audire* at the opening of this section, its reappearance a few sentences later (*urbem Syracusam maximam esse Graecarum, pulcherrimam omnium saepe audistis*, 2.4.117), and again at the close of the section, establish a parallel between Marcellus and Cicero's audience: both have heard a story; both are expected to respond. Moreover, Cicero's narrative aligns Marcellus not only with jury and readers but also with Cicero himself, who heard the stories of Verres' thefts when he was researching the case. By introducing Marcellus in this re-active role (as auditor, not perpetrator, of atrocity), the speech raises the anachronistic question of his reaction to Verres' behavior: Marcellus's response to news of an atrocity, Archimedes' death, is to be aggrieved; Cicero's response to news of Verres' atrocities is to be aggrieved; surely Marcellus's response to hearing of Verres' atrocities would be to be aggrieved. Thus by mustering up the old memory (*commemorabo*), Cicero has roused the ghost of Marcellus as a judge evaluating Verres.[49] Cicero's audience needs to consider on whose side it will stand when it enters the historical tradition, that of Marcellus and Cicero or that of Verres.

Marcellus's famous sack of Syracuse, subject of story and annals (according to the statement introducing the passage), has been embedded in Verres' plunder, which is now (according to the speech) more notorious. The tradition of Marcellus's sack, shifting and shadowy, takes distinct shape only in relation to Verres' egregious acts. This embedding distances Marcellus or any named Roman from the actual killing of Archimedes—any named Roman, that is, except Verres, the man responsible for the *direptio* of Syracuse. Thus Cicero implicates Verres in this old, old murder by creating a community out of different sets of people who perform parallel acts of listening and responding to the news that they hear: the jury, the audience of the speech, Marcellus learning of Archimedes' death.

Livy, too, when writing of the fall of Syracuse, repeatedly brings up the idea of tradition. Marcellus, says Livy, "is said" (*dicitur*) to have stood on the walls of the city and to have wept for its fate.[50] Marcellus weeps, says Livy, because he remembers the city's glorious past, and the memories that Livy puts in his head include scenes famous from the historiographical tradition.[51] When a delegation of Syracusans approaches Marcellus about saving the city, it appeals to his love of glory, his future triumph, his reputation (*fama*), and the trophies to his victory, which will stand side by side with those that Syracuse has won from Athens and Carthage (Livy 25.29.5–6).[52] As we have seen, Livy introduces the passage relating Archimedes' death with a pair of gestures toward tradition (25.31.9): it occurs, he says, among many examples (*exempla*) of anger and greed, and, he says, "it is handed down to memory" (*memoriae proditum est*). It is unclear under what moral heading this killing falls. Is it an instance of anger, greed, or both? The Latin does not tell us. Even the word *exempla* is ambiguous here, for Livy could be talking about either events or the memory of events, or he may not even be separating the two ideas.[53] Finally, after saying that Marcellus saw to Archimedes' burial, Livy goes on to say that Archimedes' name and memory protected his connections. In short, the death story is flagged as part of the system that preserved the memory of Roman *res gestae*.

We shouldn't be surprised. As Harriet Flower points out, the life of Marcellus—the "sword of Rome," warrior, and philhellene—stands at the head of Roman historiography, Roman epic, and even Roman drama. Fabius Pictor, the first Roman historian, both took part in events of the Second Punic War and wrote his history of it in order to explain Roman mores and policy to Roman, Greek, and Carthaginian.[54] The epic poet Naevius's *Bellum Punicum*, which treated the first of the Punic wars, was written during the second. The first known tragedy on a contemporary Roman topic, also by Naevius, was about Marcellus and was titled *Clastidium*.[55] Ennius, who wrote the first Latin hexameter epic, experienced the Second Punic War firsthand and included it in two books of his *Annales*. Thus the "Death of Archimedes," which concludes both Cicero's and Livy's accounts of the sack of Syracuse, takes place, like Marcellus's own death a few years later, at a point of transition, where the Roman system of aristocratic competition, one marked by an obsession with glory and commemoration, begins to take over Greek literary models and adapt them to glorifying and commemorating Romans.[56] While the death of Archimedes marks a shift in raw power (Greek machines cannot, in the end, keep Romans out; ergo Roman kills Greek), the aftermath of the death marks a shift in the power of evaluation and commemoration. At this pivotal moment, when Romans start appropriating the Greek genre of historiography, later texts present a Roman determining that the man who represents the acme of Greek mathematical genius is worthy of protection and then honorific burial.[57]

It is, admittedly, a commonplace that the most valuable honor is that which comes from one's enemy. This is true of Marcellus himself, who was buried by Hannibal, at least in some versions of the story.[58] But the death of Archimedes is especially significant because it happened at Syracuse. Archimedes died in what Cicero and Livy called the most beautiful and most ornamented of all cities. Archimedes defended his city's walls and died within them as his city suffered the consequences of its fall. Livy reports the death and burial of Archimedes as the final events in his account of the fall of Syracuse, then turns from the burial to say, "and this, for the most part, is how Syracuse was captured. . . ."[59] He presents the death as a pivotal point, both in systems of commemoration and in other ways of ordering the world: in his narrative, Archimedes' figures in the dust mark the last space under Syracusan control, and Archimedes' place of burial marks the first space under Roman control. The narrative before his death refers to the chaos in the streets in the final moment of the city's fall; the narrative after his death refers to the ritual of a funeral in the first moments of peace. Archimedes' machines once defended the city; after his death, his memory and name serve as a defense, a *praesidium,* for his connections.

Valerius Maximus is even more explicit. He says that the Roman soldier had broken into Archimedes' house, a detail that helps present the story as the breaching of concentric fortifications—first the city wall, then Archimedes' house, then his hands held protectively over his diagrams. Archimedes' ineffective plea that his diagrams not be touched stands in stark contrast with the effectiveness with which his blood confuses their lines, a vivid final detail. That the killing arises from the new world order comes across immediately in the soldier's motive for the murder, which is based on his assessment of Archimedes' behavior: he kills him "as disregarding the victor's command" (*perinde quasi neglegens imperi uictoris*). This establishment of a new order is reflected more distantly in Valerius's own act of judging Archimedes and isolating him from the other, positively represented examples of *industria.* Pliny the Elder assumes this new order when he writes that Marcellus's command that Archimedes not be killed was evidence (*testimonium*) of his genius. Marcellus's opinion is the one that matters.

Archimedes, then, becomes a symbol of this transition. His isolation, moreover, makes him representative of all Syracusans. Polybius says that Archimedes' defense of Syracuse showed what "one soul" (μία ψυχὴ) could accomplish (8.3.3). Cicero suggests that Archimedes was all that Marcellus sought. Livy explicitly refers to him as the one man (*unus homo* [Livy, 24.34.1]) who held off the Roman fleet with his machines.[60] According to Valerius, Marcellus decreed specifically that Archimedes be spared (*ut capiti illius parceretur edixit* [see n. 2]). According to Pliny, Marcellus decreed that Archimedes alone not be violated (*ne violarentur unus* [see n. 14]). Moreover, Plutarch, in his first version of the "Death of Archimedes," says explicitly that Archimedes happened to be alone when the Roman soldier came upon him. (We shall also see how Silius Italicus isolates Archimedes from the joyful Syracusan mob.) This focus on Archimedes singles him out from the other Syracusans; it has, moreover, the side effect of suggesting not just that Archimedes alone was to be saved or that he was killed alone but, in a striking revision of the truth about the bloody sack, that he alone was killed. It is no great leap from Polybius's idea of "one soul" against many or Livy's of "one man" against the Roman fleet to that of "one life for many." It requires only a shift from thinking about the physical leverage provided by Archimedes' machines to thinking about the religious leverage provided by sacrifice.[61]

If deaths attending the building or the violation of walls bear special significance, as David Konstan has argued for Livy, Archimedes' death, that of the defender of walls, can be seen as a sacrifice attending Syracuse's capture and "refoundation" as a city under Roman control.[62] In fact, Livy's "Death of Archimedes" paradoxically recalls his account of the death of Remus. Each story emphasizes the role of the attacker and defender of walls; in each, the man killed is the one who exposes the other to ridicule concerning walls— Remus by leaping them to make fun of his brother, Archimedes by toying with the Roman attackers.[63] In terms of space and structure, the events complement one another exactly, as one would expect of deaths that signify, respectively, growth from the center and growth at the margin: the first shows the eponymous Romulus, the "man of Rome," defending the first space that will become Rome; the second shows an anonymous "man of Rome" attacking the last defended space in a city being captured by Rome. Romulus kills his twin, his other self; the Roman kills a person whom he does not even know. The builder of the walls jumps alertly to their defense; the great defender of the walls is unaware of his impending death and (in Livy) does not even defend his diagrams.[64] He bends his head to the earth like a willing victim. As Livy, like Valerius, represents it, preconquest Syracuse dwindles to those diagrams in the dust. Archimedes' death is the end of old Syracuse, and his burial is the beginning of a city belonging to Rome. Cicero, too, touches on the pivotal nature of the events, when he says of Marcellus, "you would say that [Syracuse] was founded by him who captured it."[65]

Silius Italicus's *Punica* is the only surviving Roman epic that specifically includes the death of Archimedes.[66] Three hundred years after the fall of Syracuse, when even the battle over Marcellus's memory is long over, Silius represents the city's fall as part of a highly idealized past. In doing so, he brings into clear view some of the ideas implicit, or only hinted at, in the other sources and extends the range of the story's ideological potential.

Silius places the death scene at the end of book 14, in his account of the city's fall, a climactic passage that celebrates Marcellus's mercy toward the conquered.

> His tectis opibusque potitus
> Ausonius ductor postquam sublimis ab alto
> aggere despexit trepidam clangoribus urbem,
> inque suo positum nutu, stent moenia regum
> an nullos oriens uideat lux crastina muros,

ingemuit nimio iuris tantumque licere
horruit et propere reuocata militis ira
iussit stare domos, indulgens templa uetustis
incolere atque habitare deis. sic parcere uictis
pro praeda fuit, et sese contenta nec ullo
sanguine pollutis plausit Victoria pennis.[67]   (14.665–75)

[Now master of this city and this wealth,
the Ausonian general, on high, from a lofty rampart,
looked down on the city shaking with cries.
It was his decision whether the walls of kings stand
or tomorrow's rising sun see them no more;
he groaned at excess of his power
and shuddered that so much was allowed.
And, quickly restraining the anger of his soldiers,
he ordered that the houses stand
and granted to the old gods their temples to inhabit.
Thus sparing the conquered was as booty, and sufficient unto herself,
her wings untainted with blood, Victory applauded.]

Whereas, in Livy's account of the city's fall, Marcellus looked down on Syracuse and wept as he thought of its illustrious past, in Silius's account, Marcellus looked down on Syracuse, recoiled from the thought of his power over its
future, and ordered that it be spared. After pointing out how Victory herself
rejoiced in the outcome, the narrator turns to Archimedes.

Tu quoque ductoris lacrimas, memorande, tulisti,
defensor patriae, meditantem in puluere formas
nec turbatum animi tanta feriente ruina.   (14.676–78)

[You too, who must be remembered, won the tears of the general,
you defender of the fatherland, such great ruin striking you, while you
were undisturbed in mind and concentrating on your figures in the
  dust.]

The narrator goes on to say how the other Syracusans, the "remaining crowd"
(*reliquum vulgus,* 14.679), responded to Marcellus's decision—"its mind released in joy" (*resoluta in gaudia mente,* 14.679)—and that "vanquished vied
with victors" in celebration (*certarunt victi victoribus,* 14.680). Calling Marcel-

lus a rival in genius with the gods above, the narrator credits him with "founding the city by saving it" (*servando condidit urbem*, 14.681). He predicts that Syracuse will "stand as a notable trophy" (*stabitque insigne tropaeum*, 14.682) that "will allow posterity to come to know the character that generals used to have" (*et dabit antiquos ductorum noscere mores*, 14.683). Book 14 concludes with an exclamation contrasting these fortunate Syracusans with recent victims of imperial greed. Lands and seas would be plundered of riches, says the narrator, were not Domitian now protecting them from robbery (14.684–89). As in Cicero's *Verrines*, here too a memory of Marcellus's restraint provides a foil for the rapacity of the recent past to which the narrator is eyewitness.[68]

The apostrophe directed at Archimedes singles out his death as the striking exception to this generally "bloodless" victory. Yet had we not already seen how the other sources evade the death itself so as to put emphasis on Marcellus's response, even this sad event would seem oddly bloodless. The narrator does not say explicitly how—or even that—Archimedes died. He talks around the death by describing his addressee as a person "who must be remembered," referring to Marcellus's reaction by way of "the tears of the general" and then making a vague metaphorical reference to "such great ruin striking you."[69] As we shall see, this omission is meaningful, given Silius' knowledge of both Livy's account and the Roman epic tradition.

In alluding to Archimedes' death immediately after describing Marcellus's view from the rampart, Silius brings together two sources of Marcellus's sorrow: the prospect of destroying Syracuse and the news of Archimedes' death.[70] By beginning, "You too, who must be remembered, won the tears of the general," the *Punica* reverses the chronological order of the events, in which, first, Archimedes is intent on his figures; second, he is killed; third, Marcellus mourns. This reversal places emphasis on Marcellus's tears and singles out emotion as the significant common feature of the two episodes. As we shall see, this emphasis on emotion links Archimedes' death to other epic scenes that bring together tears, death, and commemoration.

Marcellus's sorrow isolates him from the joyful Syracusan mob, just as Archimedes' state of mind isolates the scientist from his fellow Syracusans. The passage draws a contrast between Archimedes' calm (he is "undisturbed in mind") and the uproar in the city (which is "shaking with cries"), between his focused concentration (he is "concentrating on . . . figures in the dust") and the wild happiness of the mob, which has its "mind released in joy." Even the metaphor of *ruina* striking suggests that, although Marcellus saved the temples, houses, and walls of Syracuse for everyone else, the fabric of the falling city collapsed on Archimedes alone.

The passage thus sets both Marcellus and Archimedes apart from the mob. Both the juxtaposition of the two scenes and the two men's isolation reinforce another parallel, one suggested by their actions: Marcellus looks down on the city from the rampart (*sublimis ab alto / aggere despexit*), while Archimedes looks down on his figures drawn in the dust (*meditantem in puluere formas*). Marcellus is Jove-like or at least perceives himself to be: he grieves that the fate of Syracuse rests on his nod (*inque suo positum nutu*); he lords it over the Syracusan gods, granting them their temples to inhabit (*indulgens templa uetustis / incolere atque habitare deis*). The image draws a parallel between Marcellus controlling his soldiers' anger and, consequently, the fate of Syracuse, on the one hand, and his maintaining and controlling cosmic order, on the other. Archimedes reproduces that order on a small scale by drawing his geometrical figures. The change in scale conveys a message: at the beginning of the siege, Greek intellect was more than equal to Roman arms (Archimedes defended Syracuse with "cunning more powerful than arms" [*astus pollentior armis*, 14.338]); now the peculiar and almost divine genius of benevolent Roman imperialism dwarfs Greek intellectual power.

By juxtaposing Marcellus weeping as he looks down from the walls and Archimedes dying as he looks down on his figures, the *Punica* brings together scenes that are already, in other authors (e.g., Cicero and Livy), rich with references to tradition and historiographical intertextuality. By juxtaposing these memory and tradition-laden scenes, Silius creates an atmosphere conducive to further remembering. Then, by commenting on and exploiting a gap in his story, he calls on readers to exercise their memories. Instead of naming Archimedes (impossible here, of course, since the name does not fit into a hexameter), the narrator addresses him as *tu . . . memorande* (you . . . who must be remembered). The word *memorande* raises the possibility of the act of remembering being performed differently according to one's relationship to the text. The narrator pronouncing this apostrophe indicates his own obligation to commemorate Archimedes. Readers, too, must remember, in order to compensate for the gap in the text. They understand Archimedes as the subject of the oblique reference by remembering him, first, via prior and outside knowledge as the person who won Marcellus's tears (e.g., in Livy or Cicero); second, as the fatherland's defender (*defensor patriae*) introduced earlier, at 14.677; and third, as the person oblivious to all except his figures in the dust, another memory available only from outside tradition (e.g., in Livy or Cicero). Thus *memorande* mobilizes intertextual memory. It prompts recollection of the old story of Marcellus's grief *as story* even before it acts as a specific intratextual prompt recalling the account of Archimedes' defense of Syracuse earlier in book 14. Readers must remember Archimedes' name first as well as external traditions about him. The apostrophe also reminds readers of the embedded nature of Archimedes' biography, that, according to tradition, whenever Marcellus's victory is remembered, Archimedes' death must be remembered as well. Moreover, by insisting that Archimedes is "to be remembered," *memorande* makes readers reenact what the tradition says was Marcellus's reaction to the death: recognition of the sad news and then efforts to compensate for it by commemoration.

In fact, it is only through remembering epic conventions that a reader otherwise ignorant of Archimedes' biography knows that this passage says he died. Marcellus grieved first for his power over a city he was about to save. According to this logic, the words "you too" (*Tu quoque*), which introduce the apostrophe, should direct a remark to something or someone else Marcellus is about to preserve, and saving Archimedes is, of course, what the tradition says Marcellus intended. If one remembers epic convention, however, it is perfectly obvious that Archimedes died, because the apostrophe calls to mind other figures addressed in like manner after their deaths (e.g., Icarus in *Aen.* 6, Caieta in *Aen.* 7).[71] Moreover, the apostrophe, the vocative *memorande,* and Marcellus's groans and tears together call to mind the one person addressed as *memorande* in Virgil's *Aeneid,* the enemy Lausus, who mourned his father's death with tears and moans (just as Marcellus wept and groaned over the fate of Syracuse) and whom Aeneas killed and then mourned.[72] Finally, they bring to mind the general association, so prominent in Virgil, of death, grief, and commemoration.[73]

Moreover, in an epic so steeped in the *Aeneid,* in a passage that so markedly assimilates Marcellus to Aeneas, founder of Rome (after all, "by saving it," Marcellus "founded the city" [*servando condidit urbem*]), it is no surprise that the apostrophe to the dead Archimedes conjures up Aeneas's tour of the underworld.[74] In fact, the apostrophe specifically invokes the famous exhortation by Anchises.

Excudent alii spirantia mollius aera
(credo equidem), uiuos ducent de marmore uultus,
orabunt causas melius, caelique meatus
describent radio et surgentia sidera dicent:
tu regere imperio populos, Romane, memento
(hae tibi erunt artes), pacisque imponere morem,
parcere subiectis et debellare superbos.
(*Aen.* 6.847–53)

[Others will hammer out bronzes more softly breathing—
Indeed I believe it so, and will draw living faces from marble.
They will plead cases better or mark out the movements
Of the heavens with a rod and describe the rising stars.
You, Roman, remember to rule peoples with your command.
These will be your arts, and to impose the way of peace,
Spare the conquered and subdue the proud.]

The words "you who must be remembered" (*Tu . . . memorande*), spoken by Silius's narrator, call to mind Anchises' command "You, Roman, remember" (*tu . . . Romane memento*). Silius's phrase *parcere uictis,* used as a substantive (*sic parcere uictis pro praeda fuit*), recalls what Anchises predicts will be one of Rome's skills: *parcere subiectis.*[75]

Anchises does not name the "others" who will be better sculptors, orators, and astronomers. The unspecified *alii* can simply indicate "Greeks" in general, or, as a completely open-ended reference, it can embrace cultural excellence on the part of any non-Romans. But nothing rules out projecting an image of Syracuse onto the outline of these "others." After all, Syracuse's art was famous, tradition credited the invention of systematic rhetoric to Syracusans, and Archimedes was a "unique watcher of the sky" (Livy 24. 34.2). Moreover, after apostrophizing the future Roman, Anchises draws attention to Marcellus.[76] The passage in the *Aeneid* thus lies open to the suggestion made by the *Punica* that it refers specifically to Syracuse.[77] In consequence, Silius's statement that Syracuse "exists, therefore, for the ages" and the prediction that it "will stand as a notable trophy and will allow posterity to come to know the character that generals used to have" shows that Marcellus did fulfill Anchises' prediction of a future in which Romans excel at the military and political arts of empire: "these will be your arts" (*hae tibi erunt artes*).[78] Yet the way by which Silius guides his readers back to this Augustan image of anticipated benevolent imperialism is through the memory of a regretted death.

While Archimedes "must be remembered" by the narrator and readers of the *Punica* and while the narrator's apostrophe brings to mind other scenes of death, tears, and commemoration, the *Punica* makes no mention of Archimedes' burial, which is referred to by Livy (and by Plutarch), or of the tomb made famous by Cicero.[79] This absence avoids closure at an otherwise conclusive point, the end of the book, the end of the war for Syracuse. The avoidance of closure is in itself not surprising for several reasons: the war for Syracuse is part of the larger conflict; there are still three books to go; Scipio, not Marcellus, is the poem's major hero, and Hannibal, not Archimedes, is the Romans' chief opponent; and even in dealing with Scipio and Hannibal, the poem in general avoids closure.[80] Yet there is one more explanation: in the *Punica*, Archimedes is not honored once and for all with a grave; instead, commemoration of him is ongoing and participatory. The poem remembers him, and it compels every reader who encounters this passage to remember him. After all, the grave was long forgotten.

The idea of poetry as commemoration works together with the representation of booty and trophies in this passage to argue that active commemoration resides not in the monuments but in the poet mobilizing and the reader responding to the many strands of tradition. Concluding his account of the fall of Syracuse, Livy pointed out the extraordinary quantity of spoils taken from the city. He returned to the topic at the end of book 25, relating their arrival in Rome and commenting on their morally pernicious influence and the precedent Marcellus set by taking them. In addition to talking around Archimedes' death, Silius, contrasting Marcellus's restraint to modern greed, deflects the issue of the famous spoils, in a manner similar to Cicero comparing Marcellus's restraint to Verres' rapacity. To be sure, Silius shows that Syracuse was extraordinarily rich in luxury items, but he lists them not when describing the sack or the triumph but when describing the city on which Marcellus cast his sorrowing gaze. Thus the ornaments and prizes—the paintings, bronzes, tapestries, cups, and pearls—stand, together with the houses, temples, and trophies of ancient kings, among the items that Marcellus would save. The trophy that Marcellus wins, the spoils he carries off from Syracuse, are purely metaphorical (*sic parcere uictis / pro praeda fuit; stabitque insigne tropaeum*). This leaves room for epic to contribute. Like Archimedes' death, Marcellus's victory is commemorated in an ongoing and participatory manner: the *Punica* celebrates it; the reader remembers that its moral perfection was predicted in the *Aeneid*.

On these terms, our Syracusan genius is as much a victor as the philhellenic Roman, since he, too, carried off a prize, Marcellus's tears (*ductoris lacrimas . . . tulisti*).[81] By representing Marcellus's tears as a prize, the passage turns Archimedes' death into a victory and presents the *insigne tropaeum*, an intact Syracuse, as a monument commemorating him as well as Marcellus. The reality of conquest—killing the enemy, taking his property, becoming a winner by making another a loser (even, as in epic or history, killing in single combat and stripping spoils)—becomes, in the realm of metaphor, an exchange of prizes. Marcellus gains glory from Syracuse, and the representative Syracusan wins tears in return. The expansion of the tradition by the addition of more layers of commemoration glosses over Roman military expansion and, unlike military conquest, multiplies prizes for both sides. The result is to erase difference. Even as vanquished and victors vie with one another in joy at

the city's salvation (*certarunt victi victoribus*), the boundaries between Syracusans and Romans disappear. By multiplying prizes and, consequently, victors,
this version of Archimedes' death places Archimedes, the "defender of the fatherland," and Marcellus, the victor who "in saving the city founded it," on the
same side. Archimedes defended the walls of Syracuse, and Marcellus saved
them, but at the end of *Punica* 14, those walls are useful only as reminders of a
lost and idealized past two removes from the present.

This is part of Silius's disturbing vision of Roman virtue.[82] On the one
hand, Marcellus's restraint helps erase difference between himself and the conquered. Even at the level of poetic commemoration, we seem to have reached
a point of equilibrium, where victor and vanquished both gain glory. On the
other hand, remembering Virgil's prediction about Rome's imperial mission
obliges Silius's readers to remember a tragic death, one that, through its use of
poetic convention, itself leads to further memories of the many deaths that
accompany conquest.

Like the "Eureka" story, the "Death of Archimedes," which seems at first glance
so simple an exemplum of *industria*, turns out to be much more complicated
when restored to its various contexts. All our sources are much later than the
event, and each has his own agenda: Cicero to make Verres look bad by making Marcellus look good and to align his own story with the history of a noble
family; Livy to commemorate the fall of Syracuse in a manner that makes it as
meaningful as Thucydides' account of the Sicilian Expedition; Plutarch to un-
derscore his subject's sensitivity to all things Greek; Silius Italicus to locate
Roman virtue—if anywhere—in the distant past. The death of Archimedes
could have been used against Marcellus. The surviving version, however, in
which an act of anger and/or greed and sheer ignorance about a Greek genius
becomes proof of the victor's respect for the victim and a paean to the brilliance of Roman governance, shows the signs of a pro-Marcellan tradition,
one that "spun" the story and co-opted Archimedes for its own ends.

Finally, the representations of Archimedes' death increasingly Romanize
him. Cicero seems to criticize Archimedes as somewhat unpatriotic, but in
Livy, Archimedes is honored with burial by a Roman, and his name and
memory protect his family, as if he represents the last of the Syracusan dead
and his surviving relations the first on the list of Marcellus's *clientela*. By addressing Archimedes as *defensor patriae*, Silius Italicus completes the ironic
transformation: Archimedes, theoretical genius and absentminded Greek,
dies like a good Roman.[83]

CHAPTER 5

THE DEFENSE OF SYRACUSE

CAST AS THE STORY OF Marcellus receiving the news, the "Death of Archimedes" plays a role in the aristocratic competition for glory in republican Rome: any thug can kill an unarmed old man, but it takes greatness of character to recognize and commemorate the virtues of an enemy who represents the acme of Greek genius. The "Death of Archimedes," like the story of the spheres, also demonstrates a shift in the power to evaluate intellect, for in commanding that Archimedes be saved and then mourning and burying him, the Roman pronounces judgment on the Syracusan. The various texts that together make up the tradition of the defense of Syracuse tell the story of the beginning of this interaction between exemplary Greek and exemplary Roman. In doing so, they appear to have neatened up the past considerably, for there are no traces of Archimedes having any direct contact with a Roman until one kills him, and there are no traces of any Roman being aware of his existence until Rome attacks Syracuse during the Second Punic War.[1] Such intact purity appears even more unreal when we recall that the siege of Syracuse took place late in Archimedes' long life, one spent for the most part in a city that was for decades a loyal Roman ally.[2] We can infer from this tidiness that the several writers who comprise the tradition have been at work here shaping this story in their own interests and that both the pristine nature of Archimedes' Hellenism and the novelty of Rome's contact with his genius are important to those interests.[3]

This chapter attends primarily to Plutarch, who tells of the defense of Syracuse in greater detail than anyone else and whose version meaningfully links several features that appear separately in the others.[4] Plutarch, in turn, provides a lens through which to view in retrospect Polybius's incomplete account and Livy's less detailed one. Although Plutarch describes the Roman siege and capture of Syracuse in his *Life of Marcellus,* the central figure in his account is not the attacking Marcellus but the defending Archimedes. Moreover, Plutarch digresses twice from the main narrative of the attack, into what I call the "Life of Archimedes," and he at one point digresses from that into a pocket history of the science of mechanics.[5]

According to Plutarch (*Marc.* 14.6–14), Marcellus sailed up to the walls of Syracuse, confident in his power and reputation, but this power and reputation failed to intimidate Archimedes or his machines. These machines, says Plutarch, beginning a digression, were "by-works" (πάρεργα), because Archimedes was more interested in pure geometry than in mechanics. There follows a history of mechanics, which traces the subject back through Archytas and Eudoxus to Plato, who was outraged that anyone would use a mechanical method of geometric proof. This is why, says Plutarch, mechanics was distinct from philosophy and categorized under the military arts. Yet, Plutarch continues, Archimedes wrote Hieron that with a given force (τῇ δοθείσῃ δυνάμει), it was possible to move any given weight (τὸ δοθὲν βάρος); he even claimed that if he had another world, he could go to it and move this one (εἶπεν ὡς, εἰ γῆν εἶχεν ἑτέραν, ἐκίνησεν ἂν ταύτην μεταβὰς εἰς ἐκείνην)—the earliest version on record of his famous expression "Give me a place to stand and I will move the world."[6] When Hieron asked for a demonstration of a great weight moved by a small force, Archimedes loaded a big ship with cargo and launched it by means of a compound pulley. Impressed, Hieron asked Archimedes to design defense engines for Syracuse.

Plutarch's narrative returns to the account of the attack with a fairly detailed description of how Archimedes' engines—catapults, beams that dropped heavy rocks, grappling irons, and small catapults called "scorpions"—warded off the Romans (15.1–17.4), until the fear that the Syracusans had felt as the Romans approached crossed over to the Romans, who became terrified at seeing anything they took to be a sign of Archimedes' presence. On recognizing that Archimedes' engines were getting the better of him, Marcellus made a joke about his situation (17.1–2) and decided to besiege the city instead of continuing to try to take it by assault. A second digression points out once again that despite his mechanical inventions, Archimedes preferred to study pure geometry; that he was impractical to the point of having to be led to the baths by his slaves; and that he wanted his tomb to display an image and inscription giving the proportions of a cylinder and its inscribed sphere (17.5–11).

The digression ends, and we learn that the Romans eventually did capture Syracuse, by guesstimating the height of the wall—who needs higher math?—and scaling it with ladders while the city's inhabitants were celebrating the festival of Artemis. Marcellus entered the city, looked down from its heights, and wept at its fate (19.1–2); the Romans plundered it, and Archimedes was killed, in any of three possible ways (19.8–11); Marcellus mourned him and honored his kin (19.12). Plutarch goes on to discuss Marcellus's just dealings with the Sicilians (20.1–7) and the cultural impact of the Syracusan art on Rome.

Of the many remarkable features of this account, two are particularly interesting for our present purposes, and they are oddly related. One is that this "Life of Archimedes" interrupts the *Life of Marcellus.*[7] The other, which I shall take up first, is the apparent inevitability of this "life," that is, the way the main elements of Archimedes' story follow causally from his statement that he could use a given force to move a given weight. The demonstration of this principle, moving the big ship, impressed Hieron, thus leading to the request for defense engines, whose performance terrified the Roman soldiers and impressed Marcellus.[8] The grand statement, the demonstration, the relationship with political power, and the practical results of that relationship follow one another as if of necessity.

Yet this plotline is not inevitable. The story does not have to be told this way. In fact, no other surviving version tells it this way. Neither Polybius nor Livy include the story of the big ship, and the writers who do include it arrange their narratives differently. In the *Punica,* Silius Italicus tells of the Roman attack and the Syracusan defense, but only at the point in the poem where Archimedes' engines have brought the assault to a standstill does he identify their designer as the man who moved a loaded ship with little effort and measured the grains of sand the world could hold (14.349–51). Proclus, in his commentary on Euclid's *Elements,* says that after Archimedes moved the ship, Hieron declared that whatever Archimedes said was to be believed (1.63). Although such a declaration could have given Archimedes the authority to design and supervise the construction of defense works, Proclus does not go on to describe the engines or report the defense of Syracuse. Nor is there any causal relationship between the three narrative elements—grand statement, big ship, and defense of Syracuse—in the *Chiliades* of the twelfth-century Byzantine scholar and poet Johannes Tzetzes, who says that he is relying on Dio, Diodorus, Pappus of Alexandria, and others.[9]

Tzetzes begins his profile of Archimedes by describing the big ship, then moves immediately to the defense of Syracuse.

Archimedes the wise, the famous mathematician
who by birth was Syracusan, an old man, a geometer
living seventy-five years, was
a man who worked out many mechanical forces.
With a triple-pullied machine and just his left hand, he both [καὶ]
dragged down to the sea a merchant-ship [καθείλκυσεν
ὁλκάδα], whose capacity was 50,000 medimni.
And [καὶ] when Marcellus, the Roman general, once
was attacking Syracuse by land and sea,
he first lifted up his ships [ἀνείλκυσεν ὁλκάδας] with his machines
    and,
raising them high against the wall of Syracuse,
cast them back down to the bottom all at once, men and all.[10]

(*Chil.* 2.103–13)

Tzetzes' Archimedes does not even make the statement "Give me a place to stand and I will move the world" until several lines later, at the conclusion of the account of the defense (129–30). To be sure, readers encounter the moving of the ship and the defense of Syracuse one after another, the two stories are linked thematically as examples of Archimedes' machines at work, and vocabulary and imagery reinforce this thematic connection. (Note the parallel construction "both . . . and" [καὶ . . . καὶ], the repetition of verbs of drawing [καθείλκυσεν, ἀνείλκυσεν], and the use of the same word for both the merchant ship used in the pulley demonstration and the Roman warships [ὁλκάδα, ὁλκάδας].) Yet these events do not, together, form a story. Tzetzes' linking of the big ship and the defense of Syracuse reflects their common identity as instances of using a given force to move a given weight. He presents as an analogical or mathematical relationship what Plutarch represents as a narrative one. This analogical relationship is important, but it does not make a causal connection.

That the story does not have to be told as Plutarch does leads one to ask why it appears this way in Plutarch. What does Plutarch's way of telling it achieve? In what follows, I first take Plutarch's story apart and examine three of its narrative elements: the story of the big ship; the account of the most famous of Archimedes' defensive weapons, the grappling iron called the Hand (Gk. χείρ, Lat. *manus*); and the joke that Marcellus makes when he sees that Archimedes' machines are getting the better of him. Comparing Plutarch's accounts to those of other authors, I reexamine how these three elements relate to one another in Plutarch, and returning to the other intriguing feature of Plutarch's narrative, the digressions, I try to assess what this "Life of Archimedes" contributes to the *Life of Marcellus*.

Taking this story of Archimedes out of the context of the *Life of Marcellus* and viewing it in the larger context of the history of science, Michel Authier sees it as the canonical "passion" narrative of the ideal scientist, divorced from the mundane, wedded to ideas, engaged in a difficult relationship with the center of political power, cast out for his calling and only later rehabilitated.[11] Viewing the Archimedes digressions in the context of Plutarch's Platonism, Phyllis Culham sees them as "a highly ideological manifesto on the proper role and status of scientist-philosophers in society and on the relative values of various sorts of scientific pursuits."[12] Although indebted to these interpretations, I want to approach the narrative differently. When we examine this "Life of Archimedes" with an eye directed less toward its science and more toward its rhetoric, we can perceive that we see the life of Archimedes only through the eyes of others, first Hieron and then Marcellus. Plutarch shows them viewing and responding to Archimedes' work, and this representation itself partakes in shaping another canonical narrative, that of the "ideal Roman" as recipient and guardian of his Hellenic inheritance.

The Big Ship

Plutarch says that Archimedes loaded with many passengers and the usual freight a three-masted merchantman, which had been dragged ashore by "the great effort of many men" (πόνῳ μεγάλῳ καὶ χειρὶ πολλῇ); then, sitting a ways off, with no great effort, but "sedately moving with his hand a certain system of compound pulleys" (ἠρέμα τῇ χειρὶ σείων ἀρχήν τινα πολυπάστου), he drew the ship toward himself smoothly (*Marc.* 14.13). Moving the ship is, indeed, an impressive feat, and Plutarch emphasizes its size and the enormity of the load it carries. Yet in Plutarch's arrangement of the narrative, to move the ship is to set in motion a miniature model of a man standing on another world and moving the earth.[13] One effect, then, of Plutarch's making a causal connection between Archimedes' statement about moving the earth and his moving the big ship is to produce an awareness of dramatic changes in scale, between man and earth, between the great weight of the ship and the slight effort needed to move it, and even between the enormity of the earth and the relative smallness of the ship.[14]

The story of the big ship is not in the earliest writers who talk about Archimedes at Syracuse; it does not appear in Polybius's account, anywhere in Cicero's surviving works, or in Livy, whose account of the attack, siege, and sack survives complete. It surfaces briefly in Silius Italicus's *Punica,* where it is marked, with the expression "they say" (*ferunt*), as traditional or fantastic: "they say that Archimedes drew up onto shore a ship piled high with rocks, and did it with a womanly right hand."[15] The story also appears in the second/third-century Athenaeus, where it is identified as a quotation from one Moschion, an otherwise unknown paradoxographer; it appears, briefly, as we have seen, in the twelfth-century Tzetzes; and a ship story bearing a family resemblance to that of Athenaeus, but not naming Archimedes, is central to a satire on the folly of human desires, called "The Ship, or The Wishes" (Πλοῖον ἢ Εὐχαί), by the second-century Lucian. Although Athenaeus's source, Moschion, is of unknown date and although Lucian is later than Plutarch, looking at their versions helps bring to light features of the story that are latent in the *Life of Marcellus.* Accordingly, let us consider Plutarch's account in the context of these other stories.

Athenaeus's description of Hieron's grain ship the *Syracosia* comes late in book 5 of the *Deipnosophists,* after detailed ekphrases of examples of excess. The speaker who relates all these extravaganzas is the jurist, poet, and musician Masurius. He quotes Polybius on the procession that began the great games of Antiochus Epiphanes (5.194c–195f), then Callixenus of Rhodes on the procession of Ptolemy Philadelphus (5.197d–203c). The latter is prefaced by a description of the elaborate viewing pavilion constructed for the event (5.196a–197c). At this banquet of learned Romans looking back to a Hellenistic past, reference to Ptolemy's expenditures on these luxuries leads, by self-conscious free association, to an observation on the number of his ships, then to descriptions of two of them: a great warship, whose dimensions indicate that it is big (5.203e–204d); and the state river barge, whose dimensions show that it is bigger (5.204d–206d).[16]

What one expects to follow in this sequence is a description of the biggest ship, and one is not disappointed. The speaker, still Masurius, introduces the *Syracosia.* He says that Archimedes was the overseer (ἐπόπτης) of this project, which Masurius read about in the treatise of a certain Moschion. He goes on to quote Moschion's long and detailed description (5.207a–209). This ekphrasis has an all-embracing quality about it: the timber for the ship came from Aetna; other materials from Italy, Iberia, and the river Rhone; "and all other things needed from all over" (καὶ τἄλλα πάντα τὰ χρειώδη πολλαχόθεν). Half the ship took six months to complete. That part was done on land (5.207b). Archimedes then launched the hull, so that work might continue with the ship afloat. The rest of the construction took another six months. Thus the process of construction both took a complete year's time and wedded the polarities of land and sea. The ship even had the entire story of the *Iliad* depicted in mosaic form on the floor of some of the cabin, as well as representations of the story on the ceilings and walls. It boasted a gymnasium, promenades, garden beds (with ivy and grapevines), and the "chamber of Aphrodite" that, as I noted in the introduction, so scandalized Guglielmo Libri. The *Syracosia* also had a library, a bathroom, stalls for horses, cabins for the stable boys, and a tank for keeping fish. A row of colossi nine feet high ran round the ship outside to support the upper weight. It had eight turrets on it, a wall with battlements and decks, and defense engines, designed by Archimedes. It even had its own court for dispensing justice (5.209a).

The narrator tells us that when there was a great deal of inquiry as to how to launch the half-built ship, Archimedes fitted out a screw and "launched it by himself through the agency of a few bodies" (μόνος αὐτὸ κατήγαγε δι' ὀλίγων σωμάτων, 5.207b). We learn, in addition, that pumping even a large quantity of bilgewater from this vessel could easily be done by one man (δι' ἑνὸς ἀνδϱὸς), because of Archimedes' screw (5.208f). The text explicitly credits Archimedes with inventing this helix, or screw windlass: "Archimedes was the first to discover the contraption of the screw windlass" (πϱῶτος δ' Ἀϱχιμήδης εὗϱε τὴν τῆς ἕλικος κατασκευήν). Although this screw windlass is not the triple pulley of Plutarch, the historical questions—whether or not Hieron built this ship as he did; whether or not Archimedes launched it or any ship; and, if he did, whether he used a triple pulley or screw windlass to do so—are not my present concern. The important point is the common one of mechanical advantage. In Plutarch, launching the big ship was a visible and comprehensible illustration of Archimedes' abstract statement that a given weight can be moved by a given force. It also was proof on a small scale of Archimedes' extreme, if concrete, claim that if there was another earth, he would go to it and move this one.

Athenaeus's *Syracosia*, however, is no small-scale model, for it dwarfs even Plutarch's heavy-laden triple-masted merchantman. Indeed, the *Syracosia* takes Archimedes' claim almost literally: not just the "lady of Syracuse," it is *a* Syracuse, with all the features of a city (walls, gymnasia, courts, defenses designed by Archimedes). In the end, says Athenaeus, this ship is too big for any harbor. It sails to Alexandria and stays there, where it is appropriately renamed the *Alexandris*. Even the epigram attributed to Archimelus, which brings the description to a climactic end, insists again on the all-embracing nature of the ship: the *Syracosia* matches in height the peaks of Aetna; it is as wide as an Aegean island; its mastheads touch the stars; the cables securing its anchors are those with which Xerxes bound the Hellespont; and so on

(5.209c–e). After this, says Masurius, it is not even worth talking about Antigonous's relatively puny sacred trireme. Thus, although Athenaeus does not explicitly present the launching of the *Syracosia* as proof of the declaration "Give me a place to stand and I will move the world," the ship, which Archimedes moves "by himself through the agency of a few bodies," goes beyond Plutarch's three-masted merchantman toward literally fulfilling that claim.

The relationship in Athenaeus between the ship and the discourse that represents it reflects this absurdity. The ship itself is too big to perform as a ship; it is a floating city. At the same time, the description of the ship, the climax of a series of ekphrases, represents ekphrasis taken to an extreme, writ disproportionately large.[17] When, having concluded his "catalog of ships," Masurius changes the subject to anticipate a question about the word used for a tripod stand in Callixenus's description of Ptolemy's procession (5.209f), this trivial question about vocabulary throws cold water on readers dazzled by the ekphrastic tour de force. Ruth Webb, who makes this point, explains, "The comment adds to the gentle humour of Athenaeus' depiction of his characters: after such a verbal display of *enargeia,* the only response they can collectively summon up is a question on a point of vocabulary."[18] Later in this chapter, I will build on Webb's observations; but note here that this "gentle humour" is generated by two abrupt changes in scale, from ekphrasis to trivial question and from cosmic ship to tripod stand.

We find some of the same comic attitude toward relative size in Lucian's "The Ship, or The Wishes" (Πλοῖον ἢ Εὐχαί). The author presents four characters assembling at the Piraeus to marvel at the enormous Egyptian grain ship the *Isis.* One character says that the shipwright told him it was 120 cubits long, over 30 wide, and 29 from deck to bottom.[19] He sketches its overall appearance and decoration with the same sense of all-embracing wholeness apparent in Athenaeus: after observing that one might compare its crew to an army for size, he rounds off his description by saying that the ship was said to carry enough grain to feed all Attica for an entire year. This enormity, the speaker continues, was kept safe by one man: "And all this a small little man, already old, kept safe, turning such great rudders with a slender tiller. For he was pointed out to me, a fellow with a receding hairline and curly hair; Heron, I think, was his name [Ἥρων, οἶμαι, τοὔνομα]" (*Nav.* 6). The enormity of the vessel and the emphasis on the smallness of the person that controls its movements—the tillerman is "small" (μικρός), a "little man" (ἀνθρωπίσκος), and even the tiller is "slender" (λεπτή)—show a certain affinity to the story of

Archimedes and the big ship. To be sure, the tillerman is not Archimedes, but he does share a name with Heron of Alexandria, the first-century CE inventor and mathematician. Most important here is that, we have a great mass controlled by one man, a little man who is old, just as Archimedes was when he defended Syracuse.[20]

All the Archimedes stories use metonomy to represent the mover of this ship. Silius, Plutarch, and Tzetzes refer specifically to the hand that sets in motion the marvelous machine: Silius says that it is a "womanly" hand; Plutarch that it is just Archimedes' hand and its sedate movement; Tzetzes that, with the help of the triple pulley, Archimedes drew the ship down to the water using only his left hand (χειρὶ λαιᾷ καὶ μόνῃ). The expressions qualifying the hand or its movement emphasize the ease with which the machines allow Archimedes to carry out his task. There is no throwing the bulk of one's weight into the attempt; only part of the body does the work. The hand sets the machine in motion; the machine extends the power of the hand.

The word χείρ itself is also a metaphor for effort. Plutarch refers to "the great effort of many men" required to draw the ship up on shore and conveys the idea of many men working together with the expression "by much hand" (χειρὶ πολλῇ). Thus he sets Archimedes' effortlessness, single-handedness, and machine on one side; the great effort, many men, and weighted ship on the other. The same metaphor appears in the abstract noun that Polybius uses when introducing Archimedes in his *Histories*. He reports the numerical superiority of the attacking Roman forces, who "expected that they would, because of their multitude of hands [διὰ τὴν πολυχειρίαν], in five days outstrip the preparations of those opposing them." Then he redeploys it in a general aphorism: they did not foresee "that at times one soul is more effective than any multitude of hands [πολυχειρίας]."[21] I will return to this piece of wisdom. For now, note that the hand stands for both the supply of available labor and the input of effort.

The story of the big ship, then, presents two related ideas. The first, as Authier has observed, is that of minimal effort producing maximum result. The second is that of the potential absurdity of situations illustrating this principle. It seems absurd that a few men, one man, one little man, one little old man, just the hand of one of the above, or a "womanly" hand moves a merchantman loaded with passengers and cargo, a ship loaded with rocks, or a ship that approaches the size and complexity—the all-encompassing nature—of the entire world.[22]

The Hand

These ideas cross over to the defense of Syracuse via more use of metaphor. Two weapons, one Roman and one Syracusan, receive particularly careful description in the sources. Telling of the attack by sea, Polybius first describes the Roman engine called the *sambuca*. Marcellus took a pair of quinquiremes, removed the oars on one side, and lashed the vessels together (oarless-sides in) to make a sort of catamaran, on which was fixed both a hinged ladder and a platform that could be raised and lowered. There were ropes with pulleys on top of the masts. Men standing at the stern would pull these ropes to raise the ladder and set the platform against the wall. When raised, the mast and ladder looked like the frame, and the ropes resembled the strings of a *sambuca* (σαμβύκη), a kind of harp.[23] The second engine, the Hand, concludes Polybius's catalog of the machines with which Archimedes met the Roman attack. There were magonels and "scorpions" and beams projecting from the walls. The last carried large stones and lumps of lead. Whenever a *sambuca* approached, the beams were made to swivel around and drop the weights on it. Against attackers approaching under cover of blinds, there were machines that fired heavy rocks and those that let down an iron hand (χεῖρα σιδηρᾶν), which would clutch at the prow of a ship, lift it, then let it drop, thus capsizing or swamping it. Polybius has far more to say about the *sambuca* than about this iron hand.[24] Livy, in contrast, gives the *sambuca* only a little more than half the space he gives the Hand and does not call it by name.[25]

Plutarch says that Marcellus sailed up to Syracuse with sixty quinquiremes and an "engine on a huge platform consisting of eight ships lashed together." We learn only later, when this huge platform is about to be destroyed by a series of enormous rocks hurled by Archimedes' machines, that it was called a *sambuca,* because of its appearance. Archimedes' weapons shot stones and missiles at the Roman land forces, and swiveling beams dumped heavy weights on ships. Archimedes also employed "iron hands" (note the plural), or grappling irons.

> [O]ther [ships] were seized at the prow by iron hands [χερσὶ σιδηραῖς], or beaks like the beaks of cranes, drawn straight up into the air, and then plunged stern foremost into the depths, or were turned round and round by means of engines within the city, and dashed upon the steep cliffs that jutted out beneath the wall, with great destruction of the fighting men on board, who perished in the wrecks. Frequently, too, a ship would be lifted out of the water into mid-air, whirled hither and

thither as it hung there, a dreadful spectacle, until its crew had been
thrown out and hurled in all directions, when it would fall empty
upon the walls, or slip away from the clutch that had held it. (*Marc.*
15.2–3)[26]

Plutarch gives the Hand (here, "the Hands"!) more attention and emphasis
relative to the *sambuca* than does Polybius and describes the effects more ex-
travagantly than do either Polybius or Livy.[27]

Silius Italicus says nothing whatsoever about the Roman machines, but
having described a Syracusan tower that burns after someone on the Roman
side shoots a flaming arrow at it (14.300–15), he turns to the Hand as a novel
example of Greek cunning.

> On their part, the wretched ships had the same fortune by sea,
> for when they drew nearer the walls and the city,
> where the harbor slaps calm waves against the walls,
> an unforeseen evil with a new kind of cunning [*astu*] terrified them.
> A beam, turned by the carpenter, its knots removed all around,
> like a ship's mast, bore the weapon of a curved hand [*manus*] affixed
>   to its end.
> This, from its position high on the wall, would
> snatch the fighting men up into the sky with talons of curved iron
>   and,
> swung back, would carry them into the city.
> That force did not limit itself to men,
> but the war-bearing beams often snatched ships when,
> driven from above, they would plant the steel
> and their gripping bites into the ship.
> And as soon as, with the iron sunk into the nearest wood,
> they had lifted the ship—wretched sight—
> they cast it back, once the chains were skillfully released,
> and with such an effort gave it headlong to the deep
> that the waves swallowed the entire ship, together with its crew.
>   (14.316–32)

Silius isolates the Hand as the most spectacular defensive weapon and emphasizes its novelty. Like Plutarch, he draws attention to the terror that it in-
stills in viewers.[28] Tzetzes, like Silius, ignores the Roman engines. He refers
only to the Roman ships and the various effects that Archimedes' weapons

had on them from various distances, starting with the Hand, then moving on to machines that hurled wagon-sized rocks and finally to the mirrors that set the Roman fleet on fire.[29]

In sum, the iron grapple receives particular attention and emphasis in the sources, especially Livy, Plutarch, Silius, and Tzetzes. One reason for this attention is its metaphorical nature: it is called a Hand because it acts as one, lifting, moving, dropping, and digging in its nails.[30] Moreover, it conjures up a memorable image, that of a hand moving a ship. Plutarch's readers have seen this image before: Archimedes seems simply to set the defense of Syracuse in motion, as if it were a single, complex machine, as operable by one man as a triple pulley drawing up a ship. Both the story of the big ship and the Hand story, then, express mechanical advantage in terms of "hand power," only the Hand moves ships directly.[31]

Both Plutarch and Silius agree that the Hand in action is a terrifying spectacle. But if we view its story as a reworking of the story of the big ship—in terms of both mechanics and imagery—we can see that it has comic potential and that this potential arises from a change in scale. Although the machine's effects are disastrous for the men on the ships, a viewer watching from a distance has a different experience: he sees what looks like a big hand directly lifting a Roman warship. From this distant point of view, the machine's action brings about a relative change in size, by which the human hand grows and the quinquireme shrinks, until the image perceived is that of a human hand holding a cup. We know about the position of such a viewer, because both Polybius and Plutarch provide one and draw attention to him and his response. That viewer is, of course, Marcellus.

The Joke

Immediately after describing the iron hand lifting the ships (8.6.6), Polybius reports Marcellus's response.

Marcellus was distressed [δυσχρησστούμενος] by the resistance put up by Archimedes, and, seeing [θεωρῶν] that those inside the city thus wore out his attacks with harm and mockery [μετὰ βλάβης καὶ χλευασμοῦ], he bore the event ill [δυσχερῶς μὲν ἔφερε τὸ συμβαῖνον]; but still, making fun of his own actions [ἐπισκώπτων τὰς αὑτοῦ πράξεις], he said that Archimedes used his ships to ladle water from the sea but that his harpists had been beaten as interlopers and hurled from the banquet in disgrace [ἔφη ταῖς μὲν ναυσὶν αὑτοῦ κυαθίζειν ἐκ θαλάττης Ἀρχιμήδη, τὰς δὲ σαμβύκας ῥαπιζομένας ὥσπερ ἐκσπόνδους μετ᾽ αἰσχύνης ἐκπεπτωκέναι].³²

Polybius focalizes this scene through Marcellus, who makes his joke upon "seeing" that his attacks are being met "with harm and mockery." The joke, then, according to Polybius, reflects Marcellus's interpretation of Archimedes' defense as "mockery." That Marcellus can joke about his own efforts even when being defeated demonstrates his ability to view this difficult situation with some mental detachment. Yet since Polybius's overall portrait of Marcellus is hostile, he may have included the joke as an inappropriate response to a grave situation.³³

Livy probably read it this way, for he says nothing about Marcellus's joke; in fact, he focalizes none of his account through Marcellus. His Romans act only as a group. They are confident, as a group, that they can take the city; and Livy presents its size and sprawl from their collective point of view: "they had no doubt that they would break into some part of a city that was so large and sprawling."³⁴ When this proves impossible, they make what Livy represents as a collective decision to resort to a siege: "and so, having taken counsel, they decided, since their every attempt was an object of derision, to abandon the attack and only by laying siege to block the enemy from entrance by land and by sea."³⁵ When the idea of joking does appear, it is entirely on Archimedes' side, almost as if to point to the absence of Marcellus's jest:³⁶ he contrives the machines by which an attack, however heavy, "might be made a mockery" (ludificeretur); the attack by sea "was baffled" (elusa est); and finally, in a phrase that recalls Marcellus's perception of mockery in Polybius, there is the collective realization on the part of the Roman leadership that "their every attempt was an object of derision [ludibrio esset]."³⁷

As one would expect in a "life" of Marcellus, Plutarch emphasizes and elaborates Marcellus's reaction. After watching Archimedes' machines batter both his land forces and his ships, by day and by night, and at close and distant range, Marcellus finally backs off and mocks his own engineers.

Marcellus, however, got away; and, mocking the technicians and mechanics with him [καὶ τοὺς σὺν ἑαυτῷ σκώπτων τεχνίτας καὶ μηχανοποιοὺς], he said, "Shall we not cease making war upon this geometrical Briareus [τὸν γεωμετρικὸν τοῦτον Βριάρεων], who uses our ships as ladles in respect to the sea [ὃς ταῖς μὲν ναυσὶν ἡμῶν κυαθίζει πρὸς τὴν θάλασσαν] but has beaten and driven off our harpists in disgrace [<τὰς δὲ σαμβύκας> ῥαπίζων μετ᾽ αἰσχύνης ἐκβέβληκε], and, casting many missiles against us all at once, outdoes the hundred-handed monsters of mythology [τοὺς δὲ μυθικοὺς ἑκατόγχειρας ὑπεραίρει]?" (*Marc.* 17.1–2)[38]

Plutarch presents the joke in direct statement, so that Marcellus speaks to readers just as to his engineers. Sharing for a moment the point of view of the audience within the text, readers see the Hand from the spatial and mental perspective of Marcellus and his crew. All parties stand back and regard this situation from a distance; and from this distance, as the joke shows, Marcellus recognizes the change in scale that makes his situation essentially absurd.[39]

Thomas Nagel has drawn attention to the use of scale as a way of expressing a kind of absurdity.

> What we say to convey the absurdity of our lives often has to do with space or time: we are tiny specks in the infinite vastness of the universe; our lives are mere instants even on a geological time scale, let alone a cosmic one; we will all be dead any minute. . . . Reflection on our minuteness and brevity appears to be intimately connected with the sense that life is meaningless; but it is not clear what the connection is.[40]

In order to contemplate this change of scale, argues Nagle, one must take a distant view of events.

> Humans have the special capacity to step back and survey themselves, and the lives to which they are committed, with that detached amazement which comes from watching an ant struggle up a heap of sand. Without developing the illusion that they are able to escape from their highly specific and idiosyncratic position, they can view it *sub specie aeternitatis*—and the view is at once sobering and comical.[41]

By its use of a change in scale, Marcellus's joke shows that he has had some insight about minuteness and enormity and that he can "step back and survey" himself and his situation. Moreover, by including a reference to Briareus and the Hundred-handers, Plutarch shows Marcellus's understanding of the principle behind Archimedes' success: one man with machines can outdo many hands. (In fact, in Plutarch, mechanics can literally allow one man to *have* many [iron] hands.) Spoken by the man who won the *spolia opima,* this insight is all the more important, for it conveys the idea that what was once meaningful about life has become meaningless. The Roman who, according to Plutarch, exemplifies physical courage and strength finds himself watching from a distance as his navy is baffled by machines, which can neither feel fear nor display courage; and he is particularly humiliated by those that ridicule the traditional idea of "hand-to-hand" combat.[42] Archimedes' machines, especially the Hand, have made a mockery of what was central to the values of the Roman aristocracy, which placed a premium on steadfast physical courage. As Nagel puts it, Marcellus sees a sight that is "at once sobering and comical."

Marcellus's detachment here is all the more evident when we set Plutarch's version of the joke against Polybius's, for their differences are suggestive. Where Polybius's version uses indirect speech, Plutarch's uses direct speech and is thus more vivid. It is also more elaborate, including, as it does, the reference to the Hundred-handers and Briareus, mythological details that both make it more markedly Hellenic and suggest that Marcellus's understanding is more complete. Polybius, however, places emphasis on Marcellus's distress (δυσχρηστούμενος / μετὰ βλάβης καὶ χλευασμοῦ / δυσχερῶς), with the result that his version of the joke is less a detached comment on mechanical advantage than a forced acknowledgment of failure. In Polybius, the joke concludes the account of Marcellus's attack by sea, after which the narrative turns to the land attack led by Appius. Plutarch's version, in contrast, comes at the end of both land and sea attacks, after which the narrative returns to Archimedes' preference for theoretical, as opposed to practical, inquiry. Thus in Plutarch the joke sums up Marcellus's view of the entire situation. Finally, Polybius seems interested in the *sambuca* for its own sake and describes it in detail, but Plutarch gives it so little emphasis relative to the Hand—barely describing it and misinterpreting Polybius's eight ships lashed together in four pairs as eight ships lashed together into one—that he appears to include it only to set up Marcellus's pun.[43]

Why would Plutarch go to so much trouble for a joke? In Plutarch and Tzetzes, Archimedes claims that given a place to stand, he can move the world. Likewise, in Plutarch and Tzetzes, we have the story of Archimedes moving a big ship as proof. Athenaeus takes this story to its logical conclusion by making the boat itself replicate the world in its all-encompassing enormity. All the sources on the siege refer to a metaphorical hand or hands lifting the attacking ships. Polybius and Plutarch report Marcellus's joke in response to these machines; and in Livy, we find Archimedes' Hand and the acknowledgment on the part of the Roman leadership that its attempts are being ridiculed. It is only in Plutarch that we find all of these elements together, causally connected. What is their cumulative effect? Archimedes makes a claim that given another world, he could go to it and move this one; he is asked to prove it, so Archimedes moves a big ship, so he moves Marcellus's ships, so Marcellus makes a joke. The joke depends on what goes before. The claim and the big ship appear to have been included for the sake of the joke. The joke, then, must be very important.

As Plutarch says in his *Life of Alexander,* minor incidents, sayings, and jests illustrate character (1.2). Alan Wardman has observed that because biography does not admit the speeches that in historiography illustrate character, Plutarch uses sayings or jests for that purpose. Marcellus's joke has accordingly been taken at face value as an indication of his Hellenism: he knows myth, exotic musical instruments, and the requirements of a symposium.[44] But the fact of Marcellus producing the joke, the mechanics behind it, and its position in the text are as telling of his character in Plutarch as are the specific "Hellenic" details. Indeed, when we replace the entire "Life of Archimedes" in the *Life of Marcellus,* we can see how the "Life of Archimedes" contributes to Marcellus's characterization and how its contribution is constructed around the joke.

At the beginning of his account of the defense of Syracuse, Plutarch refers briefly to Archimedes' machines, then digresses from the main narrative of the *Life of Marcellus* into Archimedes' preference for theoretical inquiry. Next he returns to the main narrative by describing the actions of Archimedes' machines in response to the Roman attack; then he digresses once again into a discussion of Archimedes' preference for theoretical inquiry. Sandwiched between two digressions, the account of the machines is both a return to the main story of the life of Marcellus and a digression from (or perhaps an interruption into) the "Life of Archimedes." This narrative structure reflects one of Plutarch's major points about these two men's lives. In geometrical terms, these are not simply lives running along lines that can, with a little stretch of the imagination, be called "parallel," like those of Pelopidas and Marcellus. The lives of Marcellus and Archimedes run along parallel lines at times, and I do not want to say that they "intersect" (it is better not to think of Plutarch's overall narrative as following straight lines), but the narrative of Marcellus's life and that of Archimedes' life follow along the edges of two solid shapes that dovetail.

Consider, first, the parallels between Marcellus and Archimedes. Plutarch insists that both men had dual natures. In the introduction to the *Life of Marcellus,* he points out that Marcellus excelled both in war and in the pursuit of

Hellenic learning. After describing his strength and fierceness in battle, he continues,

> As to the rest of his character, he was moderate, humane, a lover of Hellenic learning so far as to honor and wonder at those who were accomplished at it, since he was unable to pursue this and learn as much as he desired to, because of his lack of leisure.[45]

Then, introducing the "Life of Archimedes," Plutarch points out Archimedes' excellence in both practical mechanics and theoretical inquiry (14.4). Circumstances constrain both men, says Plutarch. They turn Marcellus more away from Hellenic learning than he would be otherwise inclined, and they turn Archimedes more toward the mechanical side than he would be otherwise inclined. Plutarch links these circumstances to each man's birth: like other "leading Romans" of his generation, says Plutarch, Marcellus received no respite from waging war, "because of his nobility and excellence."[46] Archimedes, too, is influenced by family connections, for it is his letter to King Hieron, whom Plutarch calls Archimedes' "relative and friend," that sets in motion his turn from pure geometry to the development of armaments for Syracuse.[47]

Other parallels between the *Life of Marcellus* and the embedded "Life of Archimedes" include a famous saying by the subject; his battle with a great opponent or opponents; a report of his death; a report of his enemy's reaction to his death; and a description of a monument, which brings closure to the "life." Because this is, after all, the *Life of Marcellus*, these features are, appropriately, muted and scaled down, as it were, in the subordinated "Life of Archimedes." Archimedes' great claim to be able to move the earth is in indirect speech, as opposed to Marcellus's joke, which is in direct speech. Archimedes fights only the Romans, by land and by sea, machine by machine. Marcellus fights the Gauls, the Syracusans, other Sicilians, and the Carthaginians. Plutarch reports three versions of Archimedes' death but does not name the authors of these variants. He gives one authoritative version of Marcellus's death, and when variants do appear (in the accounts of what happened to Marcellus's body), Plutarch cites each of them by name.[48] Plutarch tells of Archimedes' tomb and indicates vaguely the contents of his epitaph (the sphere and cylinder, 17.7); more than one monument commemorates Marcellus, and Plutarch, citing a source (Posidonius) for his information, quotes directly the epigram inscribed on one of them.[49]

It is no surprise that the narrated "Life of Archimedes" is both embedded within that of Marcellus and muted in comparison. This is the *Life of Marcellus*, after all. What are unexpected and intriguing—and, to the best of my knowledge, unnoted so far—are the overall parallel structure and the way in which Plutarch makes the two men's characters dovetail (that is, complement one another) so neatly at the defense of Syracuse. The famous sayings are carefully positioned: the description of Archimedes' involvement with mechanics begins with his writing Hieron that with a given force, it is possible to move a given object; that if there were another world, he would go to it and move this one. The discussion of mechanics ends shortly after Marcellus makes his joke about the *sambuca* and the Hand. In the demonstration of mechanical advantage, the ship provides a common unit of measurement by which the two declarations dovetail: the ship stands in for both the enormous earth and a little cup. Thus Archimedes and Marcellus speak about the same phenomemon, mechanical advantage, in ways complementary in point of view and scale. Finally, the two men come together from positions that are polar opposites in other ways as well. Plutarch's Archimedes is, even by Greek standards, unusually devoted to matters of the mind—to the point of neglecting his body: he forgets to bathe, and when he does, he draws geometrical figures in the oil on his skin (17.6). Plutarch's Marcellus embodies the Roman ideal of physical strength and courage.[50]

Most important, however, are their complementary definitions of work and play. Alan Wardman offers advice for understanding Plutarch's liking for jests.

> [W]e must take into account the opposition between what is serious (*spoudaion*) and what is play (*paidia*). Plutarch is convinced that man is revealed by what he says or does at those times when he does not seem to be engaged in serious actions, such as defending the country or persuading the assembly. He has in mind Plato's idea that men's amusements are a guide to their natures (*phuseis*).[51]

The opposition spelled out by Wardman appears when Plutarch introduces Archimedes' machines, saying, "To these he had by no means devoted himself as work worthy of his serious engagement [ἔργον ἄξιον σπουδῆς], but they were, for the most part, by-works of a geometry at play [γεωμετρίας δὲ παιζούσης ἐγεγόνει πάρεργα τὰ πλεῖστα]."[52] Now, as we have seen from the introduction to the *Life of Marcellus,* the first extended point Plutarch makes about Marcellus's character is that it is divided into serious pursuits and amusements. Marcellus has war as his serious endeavor, Hellenic learning (παιδεία) as his play. Thus the use of machines at Syracuse, which is serious to Marcellus, is play to Archimedes, and pure geometry, which is serious to Archimedes, would, as Hellenic learning, fall into the category of play for Marcellus. The two men's characters dovetail in the joke, which does not convey any high Hellenic learning but, with its reference to myth and its bilingual pun, is symbolic of Hellenic learning: it represents Marcellus at play.[53] Moreover, as we have seen, they are reflected in the vivid imagery of the Hand and ship. The structure of the text reflects this dovetailing: the interruption of Roman seriousness into Archimedes' life draws him from his theoretical inquiry and engages his playful side.

Plutarch pairs the *Life of Marcellus* with his "life" of the Theban Pelopidas. The embedded "Life of Archimedes" must somehow, then, contribute to shaping Marcellus's character in this greater context[54]—and indeed it does. At the beginning of the *Life of Pelopidas,* Plutarch discusses unnecessary risk. Starting with Cato the Elder's quip that there was a difference between a man's setting a high value on valor and his setting a low value on life, he goes on to tell of a soldier who was bold as long as he had a miserable disease and of the luxury-loving Sybarites, who thought it no surprise that Spartans did not fear death. He also notes Iphicrates's comparison of an army to a body, with the general as its head (κεφαλῇ δὲ ὁ στρατηγός, *Pel.* 2.1). Accordingly, says Plutarch, Callicrates's response was not good when, told to be careful, he declared that Sparta did not depend on one man, for, when fighting, marching, or sailing, Callicrates was one man, but when general, he was not. Better was the question of old Antigonous when he was about to engage in a sea fight and someone told him that the enemy had many more ships: "But how many ships do you count me as?" (*Pel.* 2.2) Thus the worth of a commander is great, says Plutarch, and "his first duty is to save the one who saves everything else" (πρῶτον ἔργον ἐστὶ σῴζειν τὸν ἅπαντα τἆλλα σῴζοντα, *Pel.* 2.2). This, concludes Plutarch, is his preface to his lives of Pelopidas and Marcellus, great men who fell against all reason. Plutarch reiterates this point about their deaths in the synkrisis at the end (Syn. 3–4), criticizing both men, but especially Marcellus, who allowed himself to be ambushed.

Archimedes provides a foil for this aspect of Marcellus's character as well, for he is the prime example, from Polybius on, of "the one who saves everything else." Plutarch says that while the Syracusans were the body of their defense works, Archimedes was, as it were, the soul; "he kept himself and his city unconquered as much as was in him."[55] Although the parallel is not perfect—Archimedes was not a general, and a soul is not a head—the central

point where his story and Marcellus's meet, the defense of Syracuse, reinforces Plutarch's assertion about the importance of one man and thus his
strictures against taking unnecessary risk. (This is his main point, after all, in
bringing together the lives of Pelopidas and Marcellus.) Archimedes, as
Plutarch's account makes clear, did not allow himself to be killed until after
the city had fallen and the plundering was taking place.

The "Life of Archimedes" as a whole, then, helps convey rhetorically both
Marcellus's philhellenism and his weakness. It increases Marcellus's luster in
some ways, even while implicitly supporting Plutarch's major criticism of
him: Marcellus can meet Archimedes' playfulness with his own and understands the mechanical advantage resulting from Archimedes' inventions, but
he cannot transfer the lesson of Archimedes' life to himself in a way that takes
him beyond the limits of his Roman nature. Offered a superb example of the
importance of one man, he fails to grasp this lesson about leverage. That
Plutarch's Marcellus fails this test of his Hellenism is not a new observation
(Pelling and Swain both point it out).[56] Plutarch's attitude fits the high regard
for *paideia* that characterized the Second Sophistic.[57] But what I have tried to
draw attention to is Archimedes' role in establishing metaphorical boundaries between Greeks and Romans, a nice literary parallel to the engines that
kept the attackers out of Syracuse.

Archimedes plays this role in Polybius as well. He enters historiography
when Polybius describes the Romans' mental state as they prepared to attack
Syracuse.

> [H]aving made ready their wickerwork screens, and missiles, and the
> rest of the siege material, they expected [ἤλπισαν] that they would,
> because of their numbers [διὰ τὴν πολυχειρίαν], in five days out-
> strip the preparations of those opposing them, not reckoning, how-
> ever, with the ability of Archimedes [οὐ λογισάμενοι τὴν Ἀρχιμή-
> δους δύναμιν] nor foreseeing [οὐδὲ προϊδόμενοι] that at times one
> soul [μία ψυχὴ] is more effective [ἀνυστικωτέρα] than any number
> [πολυχειρίας]. But now they came to know the [truth of the] saying
> through events themselves [πλὴν τότε δι᾽ αὐτῶν ἔγνωσαν τῶν ἔργων
> τὸ λεγόμενον]. (8.3.3)

Archimedes enters the account of the Romans' expectations negatively, as the
possessor of a force with which they did not reckon and the embodiment of
a truism that they could not comprehend until it was materially illustrated
"through events themselves."[58] The Romans "expected" a certain outcome,

"not reckoning" with Archimedes' genius "nor foreseeing" that one soul is more effective than any number, and "came to know" the truth of that saying. By directing attention toward the Romans' intellectual processes, Polybius shows that in this situation they have reached the limits of their reasoning. Archimedes' ability lay beyond it. Craige Champion has recently emphasized how reasoning power, λογίσμος, is the quality that, in Polybius, distinguishes the Hellenic from the barbarian.[59] Led by the most philhellenic Roman of them all, the Romans meet the Greek genius and are found collectively lacking in the distinguishing Hellenic trait.[60] Measured against the standard set by Syracuse, they come up short by the difference of one man—"one old man," as Polybius calls him (πρεσβύτην ἕνα Συρακοσίων, 8.7.8).

We can now recognize the "one old man" idea, with its great change in scale, as typical of the interest in size shown in Hellenistic math.[61] We can even see Archimedes' δύναμις as a pun on a mathematical term, usually translated as "square power."[62] Polybius expresses the Romans' expectations as a matter of purely mathematical reckoning: they outnumbered and therefore, after a certain amount of time, would inevitably overcome their opponents. The precision of their calculations is remarkable: they would have the advantage in exactly five days. The Romans, in short, had it all figured out. But their reckoning had not kept up with what Reviel Netz sees as a key development in Hellenistic mathematics, the element of surprise.[63] It is ironic that at the point where Polybius so vividly represents the clash of Greek and Roman intellects, the Romans might be called insufficiently "concrete" in their reckoning, for the element they had left out was the smallest, "one old man."

As Polybius describes it, Archimedes' defense of Syracuse is as much a display as a defense, because his machines illustrate what is for Polybius a universal principle, "that at times one soul is more effective than any multitude of hands." Plutarch's Marcellus grasps this principle and expresses his understanding of it through the Briareus joke; Polybius's Marcellus shows only frustration. Polybius's Archimedes is a wise man, making a point, not immediately comprehended, about the working of the cosmos for these "barbarians," much as Solon did for Croesus in Herodotus (1.32, 86).[64] Thus Polybius displays for his reader a moment in the great spectacle that is part of the Romans' education, itself part of the greater spectacle of their taking control of the known world in the short space of fifty-three years.[65]

In Polybius, the Romans collectively learn their lesson only through events; in Plutarch, Marcellus grasps it in theory but fails to profit by it, because his limited Hellenic education cannot overcome the Roman obsession with winning glory by personal displays of physical courage.[66] Nevertheless,

Plutarch manages to make this a story of successful cross-cultural education in the larger picture, for his Marcellus, while missing the lesson that could have saved him, conveys others. Plutarch closes his account of the fall of Syracuse by saying:

> Although the Romans were considered by foreigners to be terrifying when they set their hand to war and fearful to encounter in battle and had given no indications of a gentle disposition or humanitarianism or of any civic virtue whatsoever, Marcellus seems, at the time, to have been the first to show the Greeks that the Romans were most just [δικαιοτάτους]. (20.1)

Plutarch goes on to give an example of Marcellus's justice to the Sicilians and, returning to the fall of Syracuse and all the booty, notes Marcellus's claim to have taught (διδάξας) Romans the appreciation of art (21.1–6).

When we view this story as part of the greater narrative of Greco-Roman cross-cultural exchange, we can see Greeks and Romans teaching one another across a boundary marked by Archimedes. In Polybius, it is the idea of reason that creates this line; in Plutarch, it is the opposition between seriousness and play—two opposing views of *paidia*. Tidying up the defense of Syracuse, both Polybius and Plutarch use the story to represent the "'true' and ancient Roman character" at a crucial point in its development, by showing its limits.[67]

CODA TO PART TWO
Claudian on Archimedes

I HERE TAKE LEAVE of the ancient Archimedes with a glance at a text by the late antique writer Claudian, whose collection of short poems includes a fourteen-line epigram on Archimedes' sphere.[1] Of Archimedes' mechanical achievements, it is the planetary sphere, not the giant hand, that receives the most attention in the surviving sources. Credit for its prominence probably belongs to Cicero, who referred to it frequently and whose works had a strong influence on the later tradition.[2] When Ovid (*Fasti* 6.277–80) mentioned a sphere fashioned "by Syracusan art," he may have alluded to Cicero's description of a sphere in *De natura deorum*.[3] Writing near the end of the second century, Sextus Empiricus, arguing for a rational, intelligent, excellent, and eternal nature ordering the universe, invoked the sphere from the *De natura deorum*; in the third century, Lactantius did much the same.[4] We have seen that Firmicus Maternus drew on the sphere passage in the *De republica* for the introduction to his *Mathesis*. In the last quarter of the fifth century, Martianus Capella portrayed Geometry holding Archimedes' sphere and included a poetic ekphrasis of it; in the sixth century, Cassiodorus described the sphere in a letter written for the emperor Theodoric, asking Boethius to design a clock for the king of Burgundy, a machine like the sphere of the *De natura deorum*, sure to impress a barbarian viewer.[5] The sphere and the hand have in common the fact that they both record or anticipate the responses of those who see them in action, viewers who marvel at the movements of the machines and the genius of their maker: Philus's wonder in the *De republica*; that of hypothetical barbarians in the *De natura deorum*; in Sextus Empiricus, that of "we viewers" who "are exceedingly amazed when looking upon it" (σφόδρα θεωροῦντες ἐκπληττόμεθα); the Romans' fear and Marcellus's joke in Polybius and Plutarch. I want to tease out the implications of this common feature for a reading of Claudian's poem.

The epigram, *Carmina minora* 51, represents Jupiter addressing the other gods while looking into a glass ball, which in turn contains a sphere of Archimedes.

> Iuppiter in parvo cum cerneret aethera vitro,
>   risit et ad superos talia dicta dedit;
> 'hucine mortalis progressa potentia curae
>   iam meus in fragili luditur orbe labor!
> Iura poli rerumque fidem legesque deorum   5
>   ecce Syracosius transtulit arte senex.
> inclusus variis famulatur spiritus astris
>   et vivum certis motibus urget opus.
> percurrit proprium mentitus Signifer annum
>   et simulata novo Cynthia mense redit.   10
> iamque suum volvens audax industria mundum
>   gaudet et humana sidera mente regit.
> quid falso insontem tonitru Salmonea miror?
>   aemula naturae parva reperta manus.'[6]

*[When Jupiter saw the heavens contained in a small sphere of glass,
he laughed and spoke as follows to the gods above:
"To such a point has the power of mortal effort advanced,
now my work is made sport of in a breakable sphere!
The rules of the sky, the conviction of things, the laws of the gods,
behold, the old Syracusan has transformed them with art.
Shut within, the vital principle serves the different stars
and drives on the living work in fixed movements.
A fabricated sun runs through his own year,
and an imitation moon returns with the new month.
And now daring Diligence, spinning her own world,
rejoices and governs the stars with a human mind.
Why do I marvel at harmless Salmoneus with his false thunderbolt?
Nature's rival has been found to be a little hand.]*

Claudian's epigram displays several traits that are typical of his poetry: an interest in astronomy and a display of scientific erudition;[7] allusion to earlier Latin poets (here, the adjective *Syracosius* alludes to the sphere in Ovid's *Fasti*); a love of ekphrasis; the personification of abstractions; alliteration and assonance (e.g., *progressa potentia; luditur orbe labor*).[8] Some of the epigram's other features are characteristics both of Claudian's poetry and, as we can now see, of stories about Archimedes. For example, the poem does not name the maker of the sphere, identifying him instead by his age and geographical affiliation. There is also a striking contrast in scale, as the poem juxtaposes the small and large: king of gods and men versus the small glass container (*in parvo . . . vitro*); nature versus the little hand (*parva . . . manus*); and, of course, the absurdity of enclosing the outer reaches of the heavens, the *aethera*, in so tiny a space. The abstract qualities personified in the poem are those exemplified elsewhere by Archimedes: human power (*potentia*) and diligence (*industria*). Industria, who, according to Valerius, killed Archimedes, is here daring; she rejoices and "governs the stars with a human mind." As in Cicero's *De republica* and *De natura deorum* and as in Silius Italicus's *Punica*, here, too, Archimedes' power marks the extreme boundary of human ability (*hucine mortalis progressa potentia curae*). As in the *De natura deorum*, the Archimedes of the epigram works by means of art (*arte*) and intelligence (*mente*), and his product is a replica, imitative and simulated, a rival of nature (*mentitus, simulata, aemula*). As in Firmicus Maternus, here, too, Archimedes has transformed or shifted (*transtulit*) the abstract and invisible (*iura, fidem, leges*) into the realm of the visible.

Several of Claudian's other short poems offer either a look into a closed system or a view of a figure looking into one.[9] A series of epigrams, *Carmina minora* 33–39, describes hollow crystals enclosing water. One, *Carmina minora* 38, portrays a child looking into such a stone.

Dum crystalla puer contigere lubrica gaudet
    et gelidum tenero pollice versat onus,
vidit perspicuo deprensas marmore lymphas,
    dura quibus solis parcere novit hiemps,
et siccum relegens labris sitientibus orbem
    inrita quaesitis oscula fixit aquis.

*[A boy, while he rejoices at touching the slippery crystal
    and spins the icy mass with his slender thumb,*

sees the fluids caught in the clear stone,
    the only waters which harsh winter is able to spare,
    and brushing the dry sphere with his thirsting lips
    plants vain kisses on the waters he has sought.]

The crystal poems explore the simultaneous affinity and contrast between the hard substance on the outside and the liquid that it encloses, as well as the paradox of the crystal's impermeable clarity.[10] The other epigrams do not include a viewer, but in this one, the lump of crystal is a plaything for the boy, who rejoices at spinning it in his hands. Nature fools the boy, who attempts to taste the enclosed water but cannot; the crystal is icy (*gelidum*) yet does not melt; his kisses are in vain (*inrita*).

Much as the hard, dry crystal keeps the boy from the fluid water in *Carmina minora* 38, the glass surrounding Archimedes' dynamic system divides Jupiter from it in *Carmina minora* 51. But there is a significant difference between the crystal poems and the sphere poem: aside from a reference to the small and breakable (*fragili*) glass enclosing the sphere, the latter poem neither describes the physical nature of this intricate model of the universe nor mentions the material from which it was made.[11] Instead, as in the *De republica*, we find references to the dynamic nature of the machine (*inclusus... spiritus; vivum . . . opus*), to its movement, and to the interrelationships of the parts of the *mundus* to the whole. This is as it should be, for as the *De republica* makes clear, dynamic order, not external decoration, is the proper adornment of the *mundus*.

While Claudian's epigrams represent closed systems from an external point of view, this is also a trait that we can now recognize as characteristic of stories about Archimedes, both that of the display and demonstration of the sphere and that of Marcellus as spectator at Syracuse. Jupiter presents and explains the sphere to the other gods, as Gallus explained Archimedes' sphere to Rufus in the *De republica*. Also, Jupiter responds to the sight with wit, as Marcellus did at Syracuse.

Above all, Claudian's epigram emphasizes Archimedes' playfulness, for his genius makes sport (*luditur*) of Jupiter's effort (*labor*), as Archimedes made sport of the Romans in Livy (24.34.1–16). The sphere evokes from the god both laughter (*risit*) and a statement endorsing Archimedes' brilliance, just as the engines at Syracuse did from Marcellus. Claudian's poem thus takes up the humor in Polybius and Plutarch and the idea of ridicule in Livy. Here, too, a detached point of view makes the joke possible. The hubris of Salmoneus (*quid falso insontem tonitru Salmonea miror?*), a son of Aeolus who pretended to be Zeus, did not appear funny to the god: Zeus struck him with a real thunderbolt. Yet outside of Archimedes' closed system, including its *aethera*, Jupiter has, literally, another world from which to examine this one, a position from which hubris evokes only laughter. The smallness of the sphere—if it is small relative to the size of a human being, how much smaller it is relative to Jupiter—ensures that it is no threat to his sovereignty. Retrojecting this image of a laughing god onto the account of Marcellus at Syracuse (Silius Italicus, remember, compared Marcellus implicitly to Jupiter), we can reread Marcellus's joke as an expression of confidence by one who, viewing the conflict from the outside, feels secure about his role in the larger picture.

PART THREE

CHAPTER 6

PETRARCH'S ARCHIMEDES

PETRARCH INCLUDES biographical accounts of Archimedes in two early prose works, the *De viris illustribus* and the *Rerum memorandarum libri.*[1] The *De viris illustribus* includes a life of Marcellus (the *De Marco Claudio Marcello,* hereinafter *Marcellus*), which in turn includes an account of Archimedes' achievements and death.[2] The *Rerum memorandarum libri* (hereinafter *RM*), a collection of moral examples based loosely on Valerius Maximus, opens with a discussion of leisure and solitude followed by a set of examples, then turns to concepts that, says Petrarch, depend on and adorn leisure and solitude. These are study (*studium*) and learning (*doctrina*); and Petrarch includes Archimedes among the exemplary possessors of these qualities.[3] In both the *Marcellus* and the *RM,* Petrarch assembles composite biographies of Archimedes by combining Valerius Maximus's "Death of Archimedes" with material from other Latin authors.[4] Written no more than a few years apart and drawing for the most part on the same sources, the two passages relate substantially the same events: Archimedes studied the heavens, invented amazing machines, defended Syracuse, was killed by a Roman soldier when the city was sacked, and was mourned by Marcellus; Cicero discovered his long-neglected tomb. Almost all the anecdotes I have discussed in the previous chapters appear together in these two texts.[5]

The *Marcellus* and the *RM* allow us to see how one particularly acute reader saw the "Life of Archimedes" from a vantage point of over fifteen hundred years after the Syracusan's death. The story changes shape as Petrarch's two versions bring different features to the fore and leave others in the background. Their basic similarities thus provide an opportunity to study the effects achieved by an author using a given set of sources for different purposes. Moreover, Petrarch presents those sources as haphazard survivors of a mostly lost tradition. We see these remains through the eyes of a reader as self-conscious as he is acute, one who "cultivated self-disclosure to an extraordinary degree, more indeed than any previous human being of whom we have record."[6] These two texts, then, offer an opportunity to observe Petrarch using the figure of Archimedes in his own self-fashioning, a process in which the fragmentary nature of the tradition plays an important role.[7] In a small way, the Archimedes story illustrates Petrarch's relationship with the lost and only partly recovered past; and a close reading of it can add to our understanding of his response to the classical tradition.

Stephen Hinds has shown how Petrarch represents the tensions of this relationship by finding and exploiting "protoPetrarchean" moments in his sources, that is, "Roman predecessor-passages which anticipate and 'model' the Petrarchean project of connecting authors across space and time."[8] Cicero's story of finding Archimedes' tomb is one such moment—and an important one. Chapter 2 of the present study examined how Cicero's narrative of discovery suggests that Archimedes prefigures Cicero himself and how the tomb brings together the two men and their cultures but also holds them apart. The passages from the *De viris illustribus* and the *RM* present both an Archimedes and a Cicero who prefigure the narrating Petrarch. Yet even as Petrarch imitates Cicero's method of self-fashioning through confrontation with the remains of the past, he makes readers aware of unbridgeable gaps in time and space between Archimedes, Cicero, and himself. The present chapter examines how Petrarch's deployment of his classical sources for Archimedes in the *De viris illustribus* and then in the *RM* contributes to his fashioning of himself as a figure occupying a metaphorical middle ground between the distant past and posterity and filling the dual role of mourner and heir to classical antiquity.

Although Petrarch's *Marcellus* follows Livy's history for events of the Second Punic War, the narrating Petrarch shies away from relating the siege of Syracuse. A detailed account of these events, he says, would fatigue his readers. So he resorts to a *praeteritio*.[9] After listing extensively the topics that he will not treat—including the siege and capture of the city—Petrarch says that he must draw attention to two items.[10] The first is Marcellus's looking tearfully upon Syracuse and thinking of its glorious past (a version of Livy 25.24.11–14); the second is his order, in response to the Syracusans' pleas, that no free citizen be violated: "'He commanded the soldiers,' as Livy says, 'that no one harm the body of a free person; the rest would be booty'"[11] Petrarch then digresses into the "Death of Archimedes" (*Marc.* 51–55) by introducing a variant version of that decree: "Valerius adds (I do not know where he read it), that he [Marcellus] decreed by name that Archimedes be spared, although the Roman victory had been delayed by his talent and effort, which had contrived many new machines for the defense of his fatherland."[12]

Although the rest of the *Marcellus* relies almost exclusively on Livy 24–26, Petrarch's biography of Archimedes does not. After introducing his story with the comment from Livy just quoted, Petrarch closely follows Valerius. The death story is itself interrupted by a digression based on Firmicus Maternus. After continuing on with Valerius's version, Petrarch adds more—this time unattributed—material from Livy, giving Marcellus's reaction to Archimedes' death. Finally, he rounds off the story with an unattributed reference to Cicero's account of discovering the tomb.[13] The words that resume the main narrative after Petrarch has related the death and burial of Archimedes (*ad hunc modum captis Syracusis*) echo Livy's return to his main narrative after he has related that same death and burial (*hoc maxime modo Syracusae captae*).[14]

This "Death of Archimedes" is all the more striking—in its length, level of detail, and use of multiple sources—because it follows closely Petrarch's explicit refusal to exhaust his readers with a detailed narrative of events at Syracuse. The story must, then, contribute something to the biography of Marcellus that details of the siege would not. (The virtually Greekless Petrarch, after all, did not have access to Marcellus's joke.)[15] When we view it in its context as part of the *Marcellus,* the features that come into focus are the edict (to which Petrarch refers repeatedly), the mistaken violation of the edict, Marcellus's grief, and his attempt to make up for the mistake (*errorem, Marc.* 55) by seeing to Archimedes' burial and protecting his family.[16] It appears that some expansion on Archimedes' life is important because Marcellus's reputation for clemency is at stake: did his command to do no harm apply to all free persons or to Archimedes alone? If it applied to Archimedes alone, then Archimedes' character must justify his special treatment and Marcellus's grief. Archimedes' character must also explain, if not justify, his death.

Yet Petrarch develops the story of Archimedes' death beyond what is needed for illustrating Marcellus's clemency, reshaping and tweaking the sources in a way that adds another dimension to Archimedes' character. The story generally follows Valerius: Marcellus commands that Archimedes be saved; a Roman soldier entering Archimedes' house demands his name;

Archimedes cannot reply, even though he defends his diagrams (in direct speech, just as he does in Valerius); and his zeal for theoretical inquiry makes him morally responsible for his own death. But instead of listing, as Valerius does, Marcellus's reasons for wanting to save Archimedes (the pleasure he took in his wisdom; his feeling that it was glorious to save Archimedes as to capture Syracuse), Petrarch digresses into a discussion of the Sicilian *astrologus* Firmicus Maternus and wonders why Firmicus did not admit that Archimedes, too, was an expert stargazer.

> This man was indeed a noteworthy astronomer, even if Iulius Firmicius, himself too a Sicilian, calls him the greatest mechanic, either because of the jealousy that holds sway particularly among equals and neighbors or because he simply thinks so, when in either case he was truly both a great astronomer and a mechanic, the exceptional discoverer and maker of works of different kinds.[17]

By drawing attention to Firmicus's calling Archimedes only a great *mechanicus* although he was exceptional at both mechanics and astronomy—Petrarch sets up a straw man to knock down while asserting that Archimedes was interested in both heavenly and earthly subjects. Petrarch's interest in this heaven/earth antithesis continues as the passage goes on; returning to the Valerian material, he calls into question the nature of the diagrams that Archimedes was studying when he died. What Valerius called simply "diagrams" (*formas*) Petrarch calls "astrological, perchance, or geometrical diagrams" (*astrologicas forte vel geometricas formas*), that is, diagrams having to do with either heaven or earth (*Marc.* 53). By drawing attention to a side of Archimedes' character that is not evident in the brief reference to his defense of Syracuse (*Marc.* 51), Petrarch takes the "Death of Archimedes" in a new direction. I will follow this line of argument further when we turn to the *RM*.

Much about the presentation of these scenes, moreover, helps to delineate the character of the narrating Petrarch. As discussed in chapter 4 of the present study, the passages of Livy that describe Marcellus contemplating Syracuse and mourning Archimedes comment on the nature of the historiographical tradition as well as on the place of both Marcellus and Livy in that tradition: when Marcellus weeps, Livy brings out the idea of tradition with *dicitur* and the historiographical subjects of Marcellus's memories; when Archimedes dies and Marcellus buries him, Livy presents the death and burial as part of a tradition handed down to posterity, one that he himself is passing on. Petrarch, likewise, when excusing himself from narrating the siege in detail, presents himself as part of a community of writers and readers, one that includes the writers and readers of antiquity.

> These events are so many in number and kind that it is no wonder they exhausted the combatants, seeing that they both exhausted the writers and still exhaust readers today. It was my plan to relieve both my reader and myself of this chore by a brief verbal detour.[18]

A variety of details in Petrarch's "Death of Archimedes" develop this authorial personality by bringing the mechanics of scholarship to the fore. Petrarch represents Valerius as a problematic source. He uses indirect speech to relate what Valerius said about Marcellus's command (*addit Valerius . . . edixisse eum nominatim ut Archimedis capiti parceretur, Marc.* 51), thus subordinating Valerius's story to his own. Then he expresses doubts about this source: he does not know where Valerius read that Marcellus excepted Archimedes by name (*ubi lectum nescio, Marc.* 51); and he reiterates his doubts as to who, Valerius or Livy, was correct about the nature of Marcellus's order (i.e., whether the command was general or specific). After telling of Archimedes' death, Petrarch expresses his doubts once again: "and if this is as Valerius reports, for others tell it differently . . ." In contrast to the points on which he is uncertain, Petrarch (like Plutarch, whom he has not read) knows certainly (*certe, Marc.* 55) that Marcellus grieved.

Petrarch's uncertainty, while marked, is not completely surprising. After all, he has introduced Valerius in order to question the nature of Marcellus's decree. But Petrarch carries this sense of uncertainty beyond what is necessary for discussing the decree. In the midst of questioning Valerius's sources and veracity, he claims not to know Firmicus's motives for calling Archimedes only a mechanic. As we have seen, Petrarch goes on to import further uncertainty into Valerius's account by raising a question about the kind of diagrams that Archimedes was studying when he was killed.

By setting out these editorial markers, Petrarch presents readers with a very clear outline of the limits of his knowledge. He knows what Valerius, Livy, Firmicus, and Cicero say. He has gathered and read their works himself, and he proves it by citing and paraphrasing them. Yet Petrarch claims that Valerius's sources, Firmicus's motives, and Livy's veracity relative to that of Valerius lie outside of his knowledge. This problematizing, together with Petrarch's statement in the preface that he has compiled material from different sources, implies that here is the frontier of knowledge on the topic.[19] The *Marcellus* passage thus generates an image of Petrarch as scholar confronting the ancient texts, reading, comparing, and commenting on them. It shows him transcribing, as it were, the story line of Marcellus's life from the Livian model and, in the course of his transcription, copying marginalia written by Valerius, who is himself presented as a previous reader and transcriber of Livy's text.[20] The rest of the Archimedes story follows as Petrarch's gloss on Valerius.

Like Plutarch, Petrarch includes a life of Archimedes in his life of Marcellus as a way of marking out limits and emphasizes the dualism of Archimedes' character. But Petrarch recasts that dualism according to a different antithesis, that of heaven and earth. Unlike Plutarch, Petrarch does not juxtapose Archimedes' dual nature and that of Marcellus. In Plutarch, the "Life of Archimedes" dovetailed with the life of Marcellus in a way that marked the limits of Marcellus's Hellenism and thus defined his character. Petrarch's "Death of Archimedes" marks the limits of the narrator's knowledge and thus defines his own character.

When we turn to the Archimedes story in the *RM*, several similarities and a few important differences strike the eye. Petrarch's sources for the passage are, once again, easily identified: he quotes Livy on Archimedes' astronomical interests, then Cicero (*Tusc.* 1.63) on the sphere; he uses Valerius Maximus and Livy again (this time without attribution) for the death, then mentions Cicero's boast about finding the tomb.[21] Except when referring to Cicero's boast, Petrarch does not repeat techniques in presenting his sources: where he quotes in the *Marcellus*, he cites in the *RM*; and where he cites in the *Marcellus*, he quotes in the *RM*. He uses almost all the same sources as he does in the *Marcellus*, except that material from Firmicus Maternus does not appear here.

This last is one of several small but meaningful differences between the two passages. Instead of embedding a digression from Firmicus that allows him to argue that Archimedes was interested in both heavenly and earthly matters, Petrarch conveys Archimedes' dualism through the very organization of the *RM* passage: it falls into neat halves, which focus in turn on Archimedes' heavenly and earthly pursuits and conclude with artifacts representing those two sides of his life, the sphere and the tomb. With such an arrangement, there can be no doubt here, as in Petrarch's *Marcellus*, about the nature of the diagrams that Archimedes was studying when he was killed. His death takes place in the section devoted to earthly matters, so they are simply *figurae geometricae* (*RM* 1.23.6). Moreover, while the *Marcellus* presents Valerius Maximus explicitly as a source, questioning his veracity and citing him by name, the *RM* leaves the material from Valerius both unattributed and unquestioned. There is no doubt about the nature of Marcellus's decree: "he made an exception of him alone from so many thousands" (*unum de tot milibus excepit,* 1.23.6). While the *Marcellus,* like Valerius, renders Archimedes' last moments vividly by directly quoting his last words, the *RM* quotes them indirectly.[22] The direct quotes in this passage come from Livy and from Cicero, who is quoted at length. One effect of this change in presentation is to make Valerius and even Archimedes himself a shade less prominent and to make Livy and Cicero a shade more so. I will return to this point later.

Most important, in the *RM,* Petrarch's position relative to his subject and sources has changed. Instead of representing himself as a scholar handling his texts, Petrarch here represents himself and his sources using metaphors employed by classical authors when they discuss mnemonic technique: the narrative is a landscape through which he moves as researcher and recorder, and his examples of *studium* and *doctrina* and the sources for these examples are items located in that space.[23] Readers working through the text travel along with the remembering Petrarch—a manner of presentation well suited to a collection titled *Rerum memorandarum libri* (Books of Things to Be Remembered).

This technique, moreover, influences the way Petrarch situates, introduces, and organizes the Archimedes material. The collection's overall structure is derived from the list of cardinal virtues at the end of Cicero's *De inventione*: *prudentia, iustitia, fortitudo, temperantia.* The first book, on *otium* and *solitudo, studium* and *doctrina,* serves as a prelude to the whole, the "vestibule of my house of virtues," as Petrarch puts it (using another metaphor from the ancient mnemonic system).[24] Petrarch arranges his examples of leisure and solitude geographically in only a general way, by presenting first Romans then Greeks. In like manner, his survey of *studium* and *doctrina* gives first Roman then Greek examples of these qualities. In setting out his examples, however, Petrarch both increases the complexity of this geographical arrangement and draws attention to it by pointing out that he violates strict chronology by putting Rome first.[25] Petrarch admits that giving Rome priority makes it hard to find a starting point for his narrative, and he goes on to express his search for examples in spatial terms: "but I know what I shall do: I shall probe [*rimabor*] the Roman armies and survey [*lustrabo*] the forum and shall coax out [*eliciam*], either from the legions in arms or from the roar of the court, souls fond of study and given over to thinking" (*RM* 1.11.4).

Petrarch accordingly begins his examples of *studium* and *doctrina* with Roman author and military man Julius Caesar. Then, having conjured up Augustus, Varro, Cicero, Cicero's freedman Tiro, and Sallust, he moves on to Livy, "so that, having left the walls of the city, we do not at once flee from

Italy" (*ut menia urbis egressi non statim ex Italia fugiamus*, 1.18.1). Petrarch also calls Livy by his geographical attribute, *Patavinus*. After Livy, Petrarch moves on to the "neighboring" Pliny, "distant in neither time nor fatherland" (*nec etate nec patria longinquus*, 1.19.1). He uses the travel metaphor even more vividly in naming his last Roman examples, Cato the Elder, followed by the actors Roscius and Aesop: "behold, I am drawn back to Rome from the middle of my trail, as if on a charge of *maiestas* or sacrilege" (*en e calle medio Romam rursus quasi maiestatis aut sacrilegii reus retrahor*, 1.20.1).[26] His inquiries in Italy complete, Petrarch goes abroad: "but enough now about our own, I must sail far from these shores and turn my prow towards the greater sea of learned men."[27] He asks himself, "What land first meets me [*occurrit*] when I have turned away [*digresso*] from the Italian shores?"[28] Sicily is the answer, and there follows Petrarch's arrival in the harbor of Syracuse, where he pays respect to Archimedes. Having done so, our narrator sails back to Magna Graecia, which he lumps in with Greece: "Pythagoras meets me first as I cross the sea to Greece" (*in Greciam transfretanti primus Pithagoras occurrit*, 1.24.1). Once dealing with Greek examples, Petrarch largely abandons the travel metaphor.

The idea of mental travel finds its most elaborate expression and has the most thematic overlap with the object of study in the Archimedes story. Petrarch goes where he pleases, when he pleases, his mental journey unhampered by space and time: "it pleases me, then, to tie up the skiff of my discourse in the harbor of Syracuse for a moment, until, having paid my respects to Archimedes, I set off for Greece."[29] Yet as Stephen Hinds points out, Petrarch's texts set up a tension between what Petrarch does metaphorically and what he can do literally.[30] Like Archimedes, he can travel in his imagination, but he cannot literally cross the divide between the living present and the lost, classical past. His skiff (*cimba*) is the vessel appropriate to the literary situation, although a bad choice for crossing the straits of Messene and the waters of the Adriatic: it is the vessel Charon uses for ferrying souls across the river Styx.[31] Petrarch's very means of conveyance, then, both connects him to the past and emphasizes the unbridgeable gap between his living self and his long-dead examples.

Thus Petrarch's travel metaphor introduces the antithesis of the mobile mind and the restrained body that is so important throughout this passage. This antithesis allows Petrarch to draw a parallel between himself, a traveler through texts, and Archimedes, who never (according to Petrarch) went anywhere: Archimedes was born, lived, and died in Syracuse (*hic Siracusis ortus illic vixit, illic obiit*, 1.23.2). Although Archimedes was bodily "restricted by the boundaries of his fatherland" (*patrie terminis arctatum, RM* 1.23.2), "his mind, nevertheless, was shut in by no borders" (*animus tamen eius nullis circumclusus finibus,* 1.23.2). Petrarch develops this mind/body antithesis further by introducing the opposition between physical sight and mental vision: "but running with the freest thought through lands and seas and all heaven, he aimed the eyes of his intelligence where human sight could not penetrate."[32] The contrast implicit in Petrarch's metaphorical journey, between mental freedom and bodily constraint, becomes explicit here. The opposition between vision and mind reappears when Petrarch describes the artifact resulting from Archimedes' highest pursuit, the sphere, in which Archimedes made the movements of the heavenly bodies clear "not only to our minds, but even to our eyes" (*non tantum animis, sed oculis etiam nostris).*[33]

Petrarch explains that Archimedes' mental achievements transcend limitations of time as well as space, for he is still *patronus* of those who discuss celestial matters: "indeed a great part of those who dispute the loftiest and doubtful inquiries that the study of heavenly matters has broadcast on earth enjoy [*utitur,* 1.23.3] him as their patron." The sphere, moreover, would be an unbelievable discovery "if the all the cosmos were not still using it" (*nisi mundus adhuc omnis uteretur,* 1.23.4). As a result, the first half of the *RM* passage gives the impression of Archimedes' presence even in Petrarch's day. It suggests that Archimedes endures because he turned his mind to the nature of the cosmos and comprehended its eternal nature.

Archimedes' endurance is further reflected in the way Petrarch represents his own relationship to his sources. In the first half of the passage, he emphasizes their presence. Petrarch quotes directly Livy, who called Archimedes "a unique viewer of heaven and stars" (*"unicus" ut Livius ait, "spectator celi siderumque"*) and then Cicero.[34]

'Effecit idem,' ut Ciceronis utar verbis, 'quod ille qui in Timeo edificavit mundum Platonis deus, ut tarditate et celeritate dissimillimos motus una regeret conversio. Quod si in hoc mundo fieri sine deo non potest, ne in spera quidem eosdem motus Archimedes sine divino ingenio potuisset imitari.' (1.23.4)

[*"This man brought about the same thing," so may I use Cicero's words, "as did the Platonic god who constructed the world in the* Timaeus, *that a single revolution regulate movements very unlike in slowness and speed. And if, in this world, it cannot come about without god, not even in his sphere could Archimedes have imitated those same movements without genius that was divine."*]

Just as people interested in celestial matters enjoy (*utitur*) Archimedes' patronage and as the cosmos uses his discovery (*uteretur*), so too Petrarch uses (*utar*) Cicero's words. The first half of the passage, then, connects Petrarch's ability to tie up in the harbor of Syracuse and "greet" Archimedes; Archimedes' ability to travel the heavens in his thoughts; Petrarch's assertions about the endurance of Archimedes' name and sphere; and Petrarch's own quotation of ancient sources, which is made possible by his direct encounters with them. Just as Archimedes' thoughts on heavenly matters—and the products of those thoughts—are free from constraints of space and time, so, it seems at this point, is Petrarch's ability to travel among written texts, see them clearly, and reproduce direct glimpses of them for readers.

Turning to Archimedes' earthly pursuits, Petrarch marks the transition explicitly: "he was a marvelous contriver not only of heavenly but of earthly things as well" (*nec celestium modo sed et terrenorum mirus artifex*, 1.23.5).[35] This second half of the *RM* passage deals with specific "earthly things" (the machines that delayed the capture of Syracuse), then takes up the theme of "earthly things" more broadly defined: Archimedes' death, his burial, and Cicero's discovery of his tomb. Petrarch emphasizes the earthly orientation of this passage by telling us that Archimedes was "a most experienced geometer and inventor of machines" (*geometrie peritissimus repertorque machinarum*, 1.23.5). After pointing out that Archimedes' machines delayed the capture of Syracuse, Petrarch moves on to Valerius Maximus's version of events. As he did in the *Marcellus*, here, too, he tweaks this source. Not only does he reintroduce the heaven/earth antithesis by committing Archimedes, in this earthly half, to studying earthly figures, but he also reintroduces the mind/body antithesis by setting mental against physical vision. In Valerius Maximus, Archimedes' mind and eyes worked together to hold his attention on the figures in the dust: he studied them, "his mind and eyes pinned to the earth" (*animo et oculis in terra defixis*, 8.7 ext. 7). Likewise, in the *De viris illustribus*, Archimedes studies his diagrams, "entirely directed toward them in eyes and mind" (*totus in illas oculis ac mente conversus, Marc.* 53). In the *RM*, however, the mind and the senses oppose one another: "the great concern of his mind had shut his eyes and ears" (*oculos atque aures ingens animi cura concluserat*, 1.23.7). Mental freedom versus bodily constraint, mental clarity versus physical vision, sight aimed at the heavens versus gaze pinned to the earth—using these antitheses, Petrarch has deployed his material in a way that emphasizes the story's Neoplatonic dualism.

The second half of the *RM* passage shows some other small but, again, meaningful changes in presentation corresponding to the shift from Archimedes' heavenly interests to his earthly ones. Petrarch uses more material from Livy, but now he uses it without citing him; he also uses the material from Valerius, without citing him. He represents Cicero's words, but via indirect speech, and informs the reader of the source of Cicero's boast only indirectly, by pointing out that Cicero was at his estate in Tusculum (*in Tusculano suo*, 1.23.8) when he wrote it.

Thus the sphere represents the celestial themes of the first half of the *RM* passage, the tomb the earthly themes of the second half. Petrarch concludes the second half of the passage by telling of the aftermath of Archimedes' death. Marcellus mourned him and protected his family, and "to the man himself he gave the only thing he could, the honor of a tomb" (*sibi vero—quod unum supererat—honorem tribuit sepulture*, 1.23.8). Cicero found his grave.

Quam quidem multo post tempore disiectam et suis etiam civibus incognitam Marcus Tullius inter densissimos vepres se reperisse et ignorantibus indicasse commemorat in Tusculano suo scribens. Quantus homuncio, cuius semirutum bustum sparsosque cineres invenisse gloriatur romani princeps eloquii! Hec fortasse prolixius quam propositi necessitas exigebat in illius memoriam dicta sint, cuius historiam nusquam deinceps in hoc opere occursuram reor. (1.23.8–9)

[And this, indeed, a long time later, scattered into pieces and unknown even to his fellow citizens, Marcus Tullius, writing at his Tusculan estate, recalls finding among the densest undergrowth and showing to those who did not know of it. What a great little man, whose half-demolished tomb and scattered ashes the first man of Roman eloquence boasted to have discovered! These things were perhaps said at greater length than necessity demanded in memory of him, whose story I do not think shall meet me again in this work.]

The second half of the *RM* passage, in contrast to the first, fixes Archimedes in place and time and makes him and his artifacts subject to their effects. Archimedes is literally put in place by Marcellus, who sees to his burial. Much later, Cicero finds the forgotten tomb, which has been scattered apart over time. Instead of pointing out how people in general engage with Archimedes' ideas in the present and enjoy his enduring patronage, this part of the passage shows how specific people—Marcellus, Cicero, and Petrarch—cultivate the memory of the absent: Marcellus provides burial; Cicero finds the tomb and records finding it (note the words *commemorat* and *scribens*); what Petrarch has written was "said in his memory" (*in illius memoriam dicta sint*).

The commemorative cast of this section is striking, as is the way in which it telescopes the generations, aligning the projects of Marcellus, Cicero, and Petrarch in honoring body, tomb, and *historia*.

Petrarch quotes or cites a passage from the *Tusculan Disputations* in each half of this "Life of Archimedes." The quotation is from Cicero's argument at book 1.63 that the soul is divine; Archimedes' sphere provides an exemplum bolstering this point. The citation is from Cicero's heavily autobiographical and self-referential passage about the tomb, which I have already examined in detail (see chap. 2). Directly quoted, the *Tusculan Disputations* reflects the transcendent, permanent, and present nature of the part of Archimedes represented by the sphere; cited by place of origin, the same text reflects the embedded, transient, and absent nature of the part of Archimedes represented by the tomb. When Petrarch refers to Cicero's boast, he draws attention to the textuality of Cicero's expressions of regard for Archimedes, for it is this that has guaranteed Archimedes' endurance across the centuries. The text of the *Tusculan Disputations,* more than the events it relates, stands behind this passage, which commemorates Petrarch's reading of the sources, just as Cicero's boast commemorates Cicero's discovery of Archimedes' tomb.

Moreover, Petrarch casts each of the commemorative acts at the tomb as an attempt to compensate for ignorance. In order to compensate for the soldier's mistake (*errorem*), Marcellus gives the honor of burial; Cicero finds the unknown (*incognitam*) tomb and displays it to the ignorant (*ignorantibus*). Petrarch similarly presents the *RM* as a response to the ignorance and apathy of his contemporaries.

While touring the Latin authors, Petrarch frequently expresses regret at the loss of ancient works.[36] Castigating his age for its inertia (*segnities*, 1.15.5) and shameful sloth (*torpor,* 1.16.2), he tots up its crimes, beginning with the loss of so much Varro, then moving on to that of Cicero's *De republica,* which survives only in tantalizing fragments. Tiro's attempt to preserve his master's writings was partly futile, says Petrarch: "his work was good, had not the sloth of the freeborn scattered what the care of the ex-slave had gathered together."[37] The loss of Sallust's *Histories* is a disgrace, the loss of so much Livy a blot on "our age" (*Sed o quantam etatis nostre maculam!*).[38] Petrarch himself has sought the second decade "with the greatest but so far vain energy" (*summa sed hactenus inefficaci diligentia quesivi,* 1.18.2), but he fears that unless men change their character, "apathetic inertia" (*incuriosa segnities,* 1.18.3) will gradually hide Livy's genius behind a fog. Remembering Pliny's productivity produces only more shame, says Petrarch (1.19.1–2); to recollect ancient authors is to recollect the scandals and crimes of posterity, which, unable to produce works of its own, has allowed those of their ancestors to be lost and has robbed future generations of their ancestral inheritance (*avitam hereditatem*, 1.19.2). Petrarch has found no similar complaint among the ancestors, since there was no such loss, and if things go as he predicts, future generations will not even know what they have lost: "The one party possesses the works of antiquity intact, and the others have no knowledge of them. Neither has reason to grieve" (*ita apud alios integra, apud alios ignorata omnia, apud neutros lamentandi materia*, 1.19.3).

Petrarch calls attention to his unenviable position between the better past and the ignorant future. I use the spatial metaphor advisedly, because his closing words on Pliny make the analogy between space and time explicit: "And so I, who neither lack a reason for grieving nor have the consolation that comes of ignorance, placed as it were on the border of two peoples [*velut in confinio duorum populorum constitutus*], and looking at the same time forward and backward, wished to hand down to the next generations this complaint, which I did not receive from my fathers" (1.19.4).[39] The phrase *velut in confinio duorum populorum constitutus,* as Billanovich notes, alludes to a passage of Seneca (*Ep.* 70.2) emphasizing the parallel between space and time, the overlap of ideas of travel and of loss.[40] Seneca portrays his life as a journey by ship, with each age occupying its own territory and with death being the end of the voyage—a cliff or a harbor, depending on one's attitude. He identifies a period between youth and old age, which lies on the border of both (*in utriusque confinio positum*). In Seneca, then, the voyage is a metaphor for the individual life, and the landscape is an extended metaphor for the ages of man.

Petrarch, in contrast, projects Seneca's coastal image onto the life and afterlife of classical culture, with the result that the metaphor of the voyage provides a structure for organizing not Petrarch's external life but his confrontation with classical texts. He extends the idea of the border country temporally by situating himself not in his own middle age but in *a* middle age, between a better past that had not yet lost its past, on the one hand, and a future that has the consolation of ignorance (*ignorantiae solamen*, 1.19.4), on the other.[41] Petrarch even shares this consolation when it comes to Greek figures, for he does not know enough about Archimedes, or Pythagoras, or Plato to know what is lost. These figures lie too far in the past, in the terra incognita beyond a linguistic frontier.

In this collection arranged geographically and presented as the narrator's tour of the ancient Mediterranean, Archimedes occupies a middle ground between the Roman and the Greek examples, between Petrarch's knowledge and ignorance. On the near side of Archimedes are his sphere and tomb, as well as the Latin texts of Livy, Cicero, and others. On the far side are the unknown, lost, or unreadable authors (Archimedes himself, Pythagoras, Plato).

When we reembed the "Life of Archimedes" from the *RM* into this context of travel as a polyvalent metaphor for Petrarch's memory of the past, his narrative, his encounters with the dead, and his progression deeper and deeper into grief, we can see that this passage, like Cicero's story of the tomb in the *Tusculan Disputations,* establishes parallels between Archimedes' activity and Petrarch's own. Indeed, the narrative up to this point has been setting up such a parallel: the *studium* with which Petrarch searches out and gathers together the lost works of ancient authors corresponds to the *studium* with which the ancient authors produced them and the *studium* that killed Archimedes; it stands in contrast to the lack of *studium* in Petrarch's contemporaries.[42] Generally, Petrarch's zeal for knowledge reflects Archimedes' own; more specifically, Petrarch's mental journey through cultural memory reflects Archimedes' travels through the heavens. Archimedes' *oculi mentis* went freely where human sight could not and achieved a clear view of the world, a view illustrated by the sphere.[43] The first half of the "Life of Archimedes" has shown how Petrarch's mind travels across time and views the past—as it is preserved in surviving Latin texts—without mediation. In the second half, the tomb, with its atmosphere of decay, takes the place of the still-used sphere. The Latin reflects this contrast by representing the past at one more remove: the attributed material and direct quotations give way to unattributed material and an indirect quotation of Cicero's boast about the tomb.

   In bringing together the beginning and end of the *Tusculan Disputations* in a passage that sets heaven against earth and contrasts divine permanence with human transience, Petrarch appears to be imitating his Ciceronian model. We have seen that the tomb passage from the *Tusculan Disputations* sets the permanence of mathematical truths (the ratio of volumes of sphere and cylinder) against the transience of the epitaph. The *Tusculan Disputations* did this in a manner suggesting that Cicero, guided by his memory of the verses, arrived just in time to save the memory of Archimedes, even as the words were being eaten away from the inscription and brush was growing up over it.[44] Thus Cicero inserted himself into a failing system of commemoration and, by doing so, demonstrated that what was once the property of Greece now belonged rightfully to Rome.

   Yet Petrarch sees Archimedes at one more remove. Just as Cicero sees Archimedes' tomb, Petrarch sees the text of Cicero's *Tusculan Disputations.*

Just as Cicero points out the tomb to the ignorant, Petrarch presents himself as pointing out dead authors.[45] In this passage of the *RM*, however, we see Petrarch inserting himself into a system of commemoration that is not only failing but must by its very nature continue to fail, for, as he has learned to his sorrow, books are as transient as monuments. As Petrarch laments, even words preserved in books are all too vulnerable and doomed to fail.

Compare the details of Cicero's and Petrarch's descriptions of Archimedes' tomb: Cicero said that he recollected verses referring to the sphere and cylinder, saw the images of sphere and cylinder above the brush, and, having cleared the brush away, approached the monument; he was confirmed in his hunch when he saw the half-eaten lines of verse at its base. Like Cicero, Petrarch says, in both the *Marcellus* and the *RM*, that Cicero found the tomb, although it was covered with brush and unknown to the locals.[46] Neither of Petrarch's texts, however, refers to the sphere and cylinder on the monument, nor does Petrarch talk about the verses that Cicero remembered. Whereas Cicero claimed that his memory of the poetry led him to the monument, Petrarch makes the monument wordless and gives the words to Cicero and Livy. Petrarch also says that the tomb had been broken apart (*disiectum/disiectam*) by age. In addition, he refers in the *RM* to Archimedes' half-ruined (*semirutum*) pyre and to his scattered ashes (*sparsosque cineres*). Cicero's text refers neither to the tomb being broken apart nor to a pyre nor to any ashes whatsoever, scattered or not. Petrarch, then, has imported metaphors of scattering and gathering into Cicero's description of the tomb, and these metaphors recall Petrarch's description of his project in the preface to the *De viris illustribus* and his description of Tiro's work in the *RM* ("had not the sloth of the freeborn scattered what the ex-slave had gathered together"). Moreover, they show that the tomb has apparently continued to deteriorate in the time between Cicero's report of it in the *Tusculan Disputations* and Petrarch's report of Cicero's boast.

Finally, these details make the description of the tomb follow the paradigm for the transience of human affairs outlined in the Neoplatonic passages of the contemporary *Africa*, Petrarch's epic celebrating Scipio's victory over Hannibal in the final years of the Second Punic War.[47] The first two books of the epic relate the young Scipio's dream in which his father appears, tells the story of his own and his brother's deaths in Spain, surveys the Roman war dead, and predicts the destinies of Hannibal, Scipio, and Rome (2.210ff.). The passage draws on both the later Scipio's dream in Cicero's *De republica* and Aeneas's conversation with his father in the underworld in *Aeneid* 6.[48]

In the *Africa,* the father's prophecies of Rome's triumph and eventual decline evolve into a discourse on the transience of earthly things (2.432ff.). The world is small, warns Scipio the father. The only true immortality resides in the heavens, because even human attempts to preserve memory are doomed to fail.

Quod si falsa vagam delectat gloria mentem,
Aspice quid cupias: transibunt tempora, corpus
Hoc cadet et cedent indigno membra sepulcro;
Mox ruet et bustum, titulusque in marmore sectus
Occidet: hinc mortem patieris, nate, secundam.   (2.428–32)

[But if your wayward heart still would find joy
in empty glory, know what prize you seek:
the years will pass, your mortal form decay;
your limbs will live in an unworthy tomb
which in its turn will crumble, while your name
fades from the sculptured marble. Thus you'll know
a second death.]   (2.555–61, trans. Bergin and Wilson)[49]

He goes on to point out that although 2.561–62 "honors registered on worthy scrolls" last longer, they, too, must fade away, and as the generations go by, Scipio will be forgotten.

There follows the prophecy of an Etruscan-born poet, a second Ennius, Petrarch himself, who will revive the faded memory. But even this revival will last only for a while, for "books too soon die" (*iam sua mors libris aderit*).[50] A posterity that tries to preserve even these works faces the impediments of flood, fire, and tempest. In short, no mortal attempt at commemoration, including a long-enduring literary tradition, can overcome the essential transience of the sublunary world.

Libris autem morientibus ipse
Occumbans etiam; sic mors tibi tertia restat.   (2.464–65)

[—nay more: the earth itself
must die and take with it its dying scrolls;
so yet a third death you must undergo.]
                              (2.599–601, trans. Bergin and Wilson)

The death of the individual, the decay of monuments, the destruction of the world with the literature in it (notice the frequence of *mors* and its derivatives)—all these forms of death restrict the human sphere. The only way to escape these restraints is to transcend them.

> . . . annorum, nate, locorumque
> estis in angusto positi. Que cuncta videntem
> huc decet, huc animos attollere.   (2.470–72)

> *[Ye mortals, O my son,*
> *are held in narrow bonds of space and time,*
> *and one who understands this truth were wise*
> *to hither, Heavenward, lift up his thoughts.]*
> (2.606–9, trans. Bergin and Wilson)

The triple death, of body, tomb, and world with all its books, and the necessity of transcending such transience and fixing the mind on heaven—these are the same ideas around which Petrarch has constructed his Archimedes story.[51]

In the *RM,* Petrarch emphasizes the freedom of Archimedes' mind over against the limitations imposed on his body. As we have seen, Petrarch introduces Archimedes by pointing out that although he was born, lived, and died in Syracuse, his mind was not restricted to the boundaries of his fatherland. The product of this mental freedom, the sphere, has achieved a certain level of immortality, since all the world still uses it. The first half of the Archimedes passage, then, shows Archimedes modeling a "protoPetrarchean" method of coping with the problem of human mortality. The second half of the passage, however, places emphasis on decay. In referring to Archimedes' "scattered ashes," Petrarch has imported into his representation of the *Tusculan Disputations* an emphasis on the decay of the body, which was not in Cicero's text and which the *Africa* lists as the first kind of decay to which man is subject. In describing the tomb as "broken apart" by time, Petrarch shows that monuments fail and cause the second death. Here, Petrarch's attitude toward the tomb overlaps with Cicero's, though Cicero's monument fails differently: the brush grows up and the words are eaten away.

Cicero's account in the *Tusculan Disputations* is more optimistic than Petrarch's account in the *RM,* possibly for two reasons: first, Cicero was addressing a cultural crisis that he perceived as immediate; second, he had confidence in the circulation and survival of written texts—at least in contrast with inscriptions, which are embedded in specific places and thus more vulnerable to the ravages of time.[52] Petrarch's view of written texts is quite different. His experience with a fragmentary classical tradition has taught him that they are all too vulnerable, as are all material attempts at commemoration. By prophesying his own career and accomplishments in the *Africa,* Petrarch inserts himself into this tradition of ultimately futile commemoration. He is part of the posterity that tries to preserve the past but meets with obstacles, and even his collections of fragments and his writings will someday fall apart. Where Cicero placed confidence in literature, Petrarch sees only a third mortality. Faced with the bonds of mortality, what can one do but contemplate eternity? Thus the post-Platonic Archimedes, the *unicus spectator caeli siderumque* (and an older contemporary of Scipio Africanus) provides a "protoPetrarchean" model of transcendence.

Petrarch's telling and retelling of the Archimedes story suggests that he found it "good to think with." What he seems to have used it to think about was his own confrontation with classical authors. His reference to Cicero's boast in both the *De viris illustribus* and the *RM* shows that he recognized what was important about the passage in the *Tusculan Disputations:* not the tomb, but the text—the highly artificial structure of written words that obscures and replaces the monument—and the moment of cultural transmission and appropriation that the text represents.

The story of Archimedes offers Petrarch a place in which to present the positive results of his own historical and biographical research and to present himself as a researcher working at the limits of the knowable. By representing the sources as problematic, the narrator of the *Marcellus* draws attention to his role as a collector, collator, and judge of texts. Taken out of context, then, Petrarch's versions of the "Life of Archimedes" superficially resemble modern biographies, both in what they say about Archimedes and in what they say about their own construction. Yet, returning the story to its context, we can see that the figure of Petrarch shares a great deal with his classical sources. Modeling his moment of reception on Cicero's, Petrarch includes it in the *RM* in a metaphorical trip from Rome, to Sicily, to Greece that is at the same time a journey from fragmentary literary ruin to fragmentary literary ruin. Like Seneca and Livy, the narrating Petrarch travels metaphorically through time and space. Like Cicero, he inserts himself into his text as an active part of the very system his text describes; and, like Cicero, he mourns what is lost, sees what is left, then displays his finds to his contemporaries. Following Cicero's footsteps, Petrarch resurrects, after centuries of apparent neglect, the Roman's trope for the appropriation of Greek learning as a metaphor for his own recovery of his classical inheritance.

CONCLUSION

ARCHIMEDES ENTERS HISTORIOGRAPHY, via Polybius, as the possessor of a power on which the Romans did not reckon and as the embodiment of a principle that they could not foresee. The attack on Syracuse, in theory, should have worked, but the Romans' mathematical reckoning had left out the unexpected: one man and his ability to leverage his genius. The surviving anecdotal tradition, then, begins as an account of the Romans' learning something new by means of visible demonstration ("through events themselves," Polybius 8.3.3). Two and a half centuries after Polybius, Silius Italicus presents a domesticated Archimedes, one completely comprehensible on Roman terms as a "defender of the fatherland" (*defensor patriae, Punica* 14.677), who, like the denizens of Virgil's Elysian Fields, deserves commemoration for his patriotic efforts. Whereas Polybius's Archimedes was, for the Romans, something new, Silius Italicus's Archimedes belongs firmly to the past: access to his story comes through knowledge of Roman epic tradition; and references to his defense of Syracuse, even to the novelty of the Hand, serve not to illustrate a principle but only to identify him. Another twelve hundred years after Silius Italicus, Petrarch includes Archimedes in his portrayal of an only partially recuperable classical tradition, placing him on the far side of the boundary between the potentially accessible and the irretrievably lost. For Petrarch, who sorrows over the loss of the Latin authors whom he partially knows, the Greek Archimedes is both too old and too foreign to mourn.

Lost to Petrarch, who had not read Archimedes or Polybius or, indeed, Plutarch or Athenaeus, was the Syracusan's playfulness, which Netz sees emerging from Archimedes' work and which surfaces in the anecdotal tradition in the form of reactions to the demonstrations of his genius. We have seen Polybius and Plutarch introducing an element of humor by reporting Marcellus's joke. This joke, as we have seen, relies on metaphorical language, on absurdity arising from a striking change of scale, and on observation of events from a distant, or external, point of view. These elements resurface in other stories. Cicero, who, judging by his surviving witticisms, was no stranger to absurdities of scale, claims to juxtapose Archimedes, a "little man" (*homunculus, Tusc.* 5.64), to the tyrant Dionysius, in order to set up the greatest absurdity, that a man of Arpinum has to find Archimedes' tomb for the men of what was once the greatest and most distinguished city.[1] The elements are also apparent in the story of the sphere, which by its nature involves both an absurd change of scale—the entire planetary system made miniature—and an external viewer, whether Philus, some barbarian, or Jupiter. According to Livy, who appears not to approve of such playfulness, Archimedes toys with the Romans, the ridicule to which he subjects them implicitly justifying his death as a representative victim of Roman anger.

The "clever demonstrations" that, as Richard Martin says, mark the wise man give the figure of Archimedes an active role in bringing out the contrast between Greek and non-Greek in specific texts.[2] In Polybius's description of the attack on Syracuse, the Romans and Archimedes play the roles, respectively, of Croesus and Solon in Herodotus: in both histories, the wealthy and powerful barbarian cannot comprehend a universal principle—in one history, that one soul can effect more than many hands; in the other, that no man can be called happy until he is dead—until that principle is made clear "through events themselves." Such incomprehension marks for Polybius the limits of Hellenism. Indeed, marking a boundary, cultural or temporal (or both) becomes one of the functions of the anecdotal Archimedes: Cicero, as we have seen, presents Archimedes' tomb as a place where he attempts to overcome the divide between the living and the dead, all the while showing how he and Archimedes are kept apart. Indeed, for Cicero, the divide between "practical" mathematics—measuring and reckoning—and geometry literally marks a boundary line between Roman and Greek.[3] Plutarch organizes the *Life of Marcellus* so as to draw a line between Greek and Roman, and Petrarch makes his references to Archimedes in ways that reinforce the boundaries he places around his own knowledge and around what he considers knowable.

The present study has examined how various audiences—viewers, auditors, and readers—have responded to Archimedes' exploits and to stories about him. Polybius presented the Romans' encounter with Archimedes as a learning experience. Cicero, our earliest source on Archimedes' death, presents the story as one of Marcellus's receiving the news. Cicero, moreover, embeds the story in his own version of the sack of Syracuse, one that relies on knowledge of historical tradition. Vitruvius, too, invokes tradition in telling the "Eureka" story. Likewise, Plutarch says that "we are not to disbelieve" (*Marc.* 17.6) the stories we hear about Archimedes' absentmindedness; Silius Italicus invokes poetic memory in place of giving Archimedes' name; and Tzetzes says that Dio, Diodorus, and many others tell Archimedes' story.

These references to tradition appear first in Cicero. So, too, it is Cicero who first recasts the stories of the great inventor as those of rediscovery and appropriation; Petrarch follows suit, adapting Cicero's stories of rediscovery into his narrative of his own fight against ignorance and time. Moreover, it is Cicero who incorporates Archimedes' technology into his own program of creating an aristocracy of Romans linked not by noble ancestors but by intellectual achievement. Apparently, much of the anecdotal tradition exists because Cicero exploited Archimedes' exemplary potential. In addition to describing the spheres and the tomb, Cicero makes several other brief references to Archimedes: in the *Pro Cluentio,* he incorporates Archimedes into a joke about calculating bribes; in his letters, he refers to a politically touchy situation as a "problem for Archimedes"; in the philosophical dialogues, he uses Archimedes and his sphere as both exemplum and emblem of an ordered world.[4] Plutarch, Lactantius, Sextus Empiricus, and Petrarch all follow Cicero's lead.[5]

Archimedes, absentminded even unto death, may have appeared in a part of Polybius now lost; in the surviving tradition, he appears first in Cicero, where, arguing for the seductiveness of study, Cicero implicitly compares him to Odysseus listening to the Sirens' song. Valerius Maximus makes the scene into the topic of an exemplum. Vitruvius invokes his abstraction in the "Eureka" story; Plutarch says that mathematics has a music like the Sirens' song, then includes the "Eureka" story among his examples of absentmindedness. The sorrowing Marcellus of Livy, Valerius Maximus, Silius, Plutarch, and Tzetzes likewise surfaces first in Cicero. It is ironic that Cicero, whose surviving work shows little evidence of his having read the works of Archimedes, did more than anyone to create him as a literary figure.

One reason Archimedes was so important to Cicero was that his spheres and his tomb were tangible links to the great Marcellus. The authors who mention Marcellus in connection with Archimedes, from Polybius to Petrarch, place emphasis on him as Archimedes' appreciative audience. Cicero develops this idea in his presentation of the spheres in the *De republica,* where the artifacts attract viewers and thus create communities of those capable of appreciating each. (Recollect the criticism, voiced by Livy and reported by Plutarch, about Marcellus's bringing the Syracusan art to Rome: it created community—groups of chattering Roman art historians.)[6] By attaching himself, even indirectly, to Archimedes' artifacts, the spheres and the tomb, Cicero could insert himself into the family tradition of the Marcelli and, consequently, graft the new intellectual elite onto the stock of the old (as-intellectual-as-it-had-time-to-be) nobility. Moreover, it allowed him to claim a dual identity as both a Roman Archimedes—defender of his *patria* with words, not machines—and a new Marcellus, a philhellene transferring to Rome the artifacts of the culture that he loved, for the betterment of Greece as well as Rome. In sum, Cicero probably latched onto Archimedes' exemplary potential because he was so prominent in shaping Marcellus's contested posthumous reputation.

Cicero's most important contribution to the anecdotal tradition, however, was to present Archimedes as a figure for loss and recovery and for the transmission of cultural capital. The rediscovery of the tomb, in particular—which sets up both Cicero's encounter with Greek knowledge as a narrative of loss and recovery and the appropriation of neglected cultural capital by a worthier heir—would become for Petrarch a way of shaping his and his generation's encounter with their inheritance from classical antiquity. If Archimedes was important to Cicero because he was important to Marcellus, he was important to Petrarch because he was important to Cicero. For Petrarch, Archimedes, his sphere, and his grave are together emblematic of a favorite text, the *Tusculan Disputations*.

Cicero, moreover, transferred emphasis from Archimedes' discoveries and illustrations of them to his own discovery of ideas and his own illustration, communication, and transmission of them. Whereas Polybius presented Archimedes' defense of Syracuse as the demonstration of a principle, Cicero presented the spheres as an illustration of the transfer of cultural capital from one people to another and from one generation to the next. Other texts reflect this transference. Vitruvius offered up the "Eureka" story as the forerunner of his own, that of the humble man serving his leader, elevating himself and his profession by paying more attention to intellectual achievement than to monetary gain. Plutarch uses Archimedes' life story, including accounts of his discoveries and defense of Syracuse, as a way of adding another dimension to his biography of Marcellus and thus illustrating his idea of the interrelationship between Greek and Roman in the late third century BCE. Most of all, Petrarch, as a reader of Cicero, develops the idea of Archimedes' deteriorating tomb as a reflection of his own encounter with a constantly vanishing past, and he develops the idea of the sphere as a reflection of the permanence of immaterial ideas.

This transfer of attention away from Archimedes and his discoveries to the tradition about them continued. Two centuries after Petrarch, in his preface to the 1569 editio princeps of Nonnus's *Dionysiaca,* Gerard Falkenburg compared Johannes Sambucus's publication of the poem favorably to Cicero's discovery of Archimedes' tomb: all Cicero did was show the Syracusans where Archimedes was buried, whereas Sambucus restored to the community of letters a long-lost source of eloquence and poesy.[7] Falkenburg's comparison alludes primarily to Cicero's telling of the anecdote in the *Tusculan Disputations* and only secondarily to the act of discovery. Falkenburg says that if Cicero could boast about finding the tomb, then Sambucus will be able to boast more justly about bringing the *Dionysiaca* to light.[8] Falkenburg's Latin makes clear his reliance on the *Tusculan Disputations*: he describes the tomb as *vepribus undique et dumetis obductum* (obstructed on all sides by thornbushes and thickets), a close paraphrase of Cicero's *saeptum undique et vestitum vepribus et dumetis* (surrounded on all sides and overgrown with thorn-bushes and thickets); in both passages, the Syracusans were ignorant about the monument to their *acutissimus,* or "sharpest," citizen.[9]

The shift from Cicero's emphasis on the novelties of Archimedes' own technological innovations to Petrarch's emphasis on changes in the technology of recording and transmitting knowledge, a shift also evident in Falkenburg, who is, after all, celebrating a first printed edition, brings us back to Heiberg and to the most recent narrative of loss and recovery associated with Archimedes. It is a story of this last century. In a 1907 *Hermes* article, Heiberg wrote of a new discovery. While completing his second edition of Archimedes' works, he had been informed by one Professor Schöne of a recently published manuscript of a prayer book, which was said by its editor to contain a palimpsest of what appeared to be a mathematical text.[10] The editor of the prayer book had quoted some of the underlying text to convey an idea of its contents. Heiberg recognized them as Archimedean. The codex in question resided in the Library of the Metochion (daughter house) of the Cloister of the All-Holy Sepulcher in Constantinople. After attempts to have the text brought to Copenhagen by diplomatic means failed, Heiberg traveled to Constantinople during the summer of 1906 and was allowed to study the manuscript, a quarto volume of 185 pages—the first 177 of parchment and containing a prayer book written in the twelfth or thirteenth century, the last pages of sixteenth-century paper. On those 177 parchment pages, says Heiberg, there appeared a palimpsest, written in "attractive light-brown ink," which had been only washed, not scraped off, and had thus survived to various degrees of legibility. Recognizing that the underlying text was, indeed, by Archimedes, he collated and transcribed as much of it as he could and took photographs of it as well. Heiberg explains this all soberly in the *Hermes* article that described the palimpsest and included the first publication of the new readings.[11] He admits his excitement only in his Latin preface to his second edition of Archimedes, where he says that it was with the greatest amazement (*cum summo stupore*) that he identified the nature of the underlying text.[12]

Heiberg produced his new edition (1910–15) from photos of the palimpsest, which contained a better text for the work *On Floating Bodies*, the bulk of a totally new text called *The Method*, and the fragment of another, the *Stomachion*.[13] In fact, the palimpsest contains parts of at least seven texts of Archimedes.[14] At some point after Heiberg saw it, the manuscript disappeared (it was first discovered missing in 1922). It was rumored to be in private hands in Paris. In 1998, it resurfaced, with extensive damage, in New York, where it was sold, in 1999, for two million dollars. The buyer then deposited the manuscript with the Walters Art Museum in Baltimore, where it resides today, undergoing study with the most sophisticated imaging devices available. It was featured in the television series *Nova* in 2003.

In *The Archimedes Codex*, Reviel Netz and William Noel, manuscript curator at the Walters Art Museum, tell the story of the palimpsest in riveting detail. They reconstruct the history of Archimedes' words and diagrams from the creation of the manuscript to the disappearance of the codex after Heiberg's visit; they tell of its further adventures from its sale in New York to its arrival at the museum; they report the progress made in preserving and studying it; and they point out some the contributions it has already made to the study of Greek mathematics. As Netz observes, however, the complete story of the palimpsest's adventures between its disappearance after Heiberg's collation and its reappearance on the auction block is yet to be told. But in Heiberg's *Hermes* article, we already have one complete narrative of loss and recovery; in fact, we have had it for two thousand years. We recognize in Heiberg's account of the palimpsest—the scholar's journey in search of half-erased words matching those already known, his rediscovery and recognition of his find (words and illustrations together!), and his publication of it—as a retelling of Cicero's account of finding the lost tomb. Likewise, Cicero's story of the spheres (itself surviving only in a palimpsest) anticipated the story of the later adventures of the Archimedes codex, that of a scholarly work transferred from the Greek world to a new "imperial" center (the United States) and given new status as a prestige object and an object of scholarship.

From papyrus, to photograph, to printed edition, to television, to website—as the adventures of the Archimedes palimpsest show, the story of Archimedes as a narrative of loss and recovery continues today, but by way of different media and in a cultural context in some ways similar to and in others very different from that of the ancients. Like Cicero, Petrarch would find familiar the trajectory of a narrative that reported the rediscovery of words obscured by time and ignorance. He might have viewed the palimpsest as yet another illustration of the frailty of the written word ("books too soon die"), and he might have considered its assumption into ethereal form, via television and the Internet, as just another way station on the inevitable journey to oblivion. Marcellus, Polybius, Cicero, and Firmicus would all understand perfectly the importance of entrusting the Archimedes palimpsest to the Walters Art Museum, a "sacred" civic space, far from the palimpsests' origins but near the center of imperial and financial power. Cicero would also approve of its exposure, from a distance, to the general public and, at closer hand, to the intellectual elite, who are allowed to manipulate the artifact and explain it to others. He would be utterly baffled, however, by the values of a culture in which the owner of an item of such intellectual importance and cultural prestige would desire to remain unnamed.

NOTES

PREFACE AND ACKNOWLEDGMENTS

1. Licinus Porcius wrote *Poenico bello secundo Musa pinnato gradu / intulit se belli-cosam in Romuli gentem feram* (cited in Gellius *N.A.* 17.21.45). What kind of Muse he meant is not clear. See Edward Courtney, ed., *The Fragmentary Latin Poets* (Oxford: Oxford University Press, 1993), 83–86.

2. Emma Dench, *From Barbarians to New Men: Greek, Roman, and Modern Percep-tions of Peoples of the Central Appenines* (Oxford: Oxford University Press, 1995), 30. See also Dennis Feeney, "The Beginnings of a Literature in Latin," *JRS* 95 (2005): 226–40.

INTRODUCTION

1. Johan Ludvig Heiberg, *Quaestiones Archimedeae* (Copenhagen: Hauniae, 1879), 4–9.

2. Heiberg, *Quaestiones,* 4 n. 1. Heiberg points out that Heracleides may be the friend who helped relay books to Dositheos. It is interesting to consider, in contrast, the figure of Pythagoras, who, because of the community that formed around his teachings, boasts three—although late and fantastic—lives. See Charles H. Kahn, *Pythagoras and the Pythagoreans: A Brief History* (Indianapolis: Hackett, 2001), 5–22.

3. See especially Arnaldo Momigliano, *The Development of Greek Biography,* 2nd ed. (Cambridge, MA: Harvard University Press, 1993). Momigliano observes: "Hellenistic biography was far more elaborately erudite than any previous biographical composi-tion. It was also far more curious about details, anecdotes, witticisms, and eccentrici-ties" (120). On ancient biography's interest in character rather than event, see S. Aver-intsev, "From Biography to Hagiography," in *Mapping Lives: The Uses of Biography,* ed. Peter France and William St. Clair (Oxford and New York: Oxford University Press, 2002), 19–36. On the "Eureka" story, see chap. 1 in the present study; on "Give me a place to stand . . . ," see chap. 5.

4. Heiberg, *Quaestiones*, 5: *quod quidam narrant, eum in Hispania quoque esse pro-fectum, id totum commenticium est* (what some write, that he also traveled to Spain, is entirely false).

5. Heiberg draws on earlier lists of ancient references to Archimedes: the preface to D. Rivault's seventeenth-century edition of Archimedes and the compilation of G. M. Mazzuchelli, *Notizie Istoriche e Critiche Intorno alla Vita, alla Invenzioni ed agli Scritti di Archimede Siracusano* (Brescia: G. M. Rizzardi, 1737). F. Hultsch's encyclopedia entry ("Archimedes," *RE* 2.1 [1895]: 507–39), which relies on Heiberg, first presents a narrative that is very similar, but without reference to sources, then catalogs and evalu-ates those sources. The narrative biographies included by other editors of Archimedes are, like Heiberg's, careful. See T. L. Heath, ed., *The Works of Archimedes* (Cambridge: Cambridge University Press, 1897; supplement, 1912; reprint, Mineolo, NY: Dover, 2002); E. J. Dijksterhuis, *Archimedes*, trans. C. Dikshoorn (Copenhagen: E. Muskgaard, 1956; reprint, Princeton: Princeton University Press, 1987); Paul Ver Eecke, *Les Oeuvres Complètes d'Archimède*, 2nd edition, 2 vols. (Paris: A. Blanchard, 1961).

6. In his Loeb edition of *Greek Mathematical Works* (Cambridge, MA: Harvard University Press, 1941; reprinted with revisions, 1993), Ivor Bulmer-Thomas does not offer readers a narrative biography but, instead, simply prefaces his translation of the *Sphere and Cylinder* (in vol. 2, *Aristarchus to Pappus of Alexandria*) with lengthy quota-tions from Tzetzes, Plutarch, Pappus, Diodorus Siculus, and Vitruvius. He adds com-mentary and caveats in footnotes. Charles Mugler begins the preface to his Budé edition with a brief narrative biography, but instead of citing sources, he simply points out that the details come from "une tradition souvent incertaine" (Charles Mugler, ed. and trans., *Archimède, Oeuvres*, vol. 1 [Paris: Les Belles Lettres, 2003], vii–viii).

7. Reviel Netz, *The Works of Archimedes Translated into English*, vol. 1, *The Two Books on the Sphere and the Cylinder* (Cambridge: Cambridge University Press, 2004), 10–11.

8. Other extravagant stories attach themselves to Archimedes: Tertullian credits him with inventing the organ (*De anima* 14.4), an attribution contradicted by Pliny (*N.H.* 7.125), Vitruvius (10.7), and Athenaeus (4.174d); see Heiberg, *Quaestiones*, 43. Pe-trarch (*De remediis utriusque fortunae*, dialogue 99) reports that some credit him with inventing the cannon: see M. Clagett, *Archimedes in the Middle Ages*, vol. 3, *The Fate of the Medieval Archimedes* (Philadelphia: American Philosophical Society, 1978), 1340; D. L. Simms "Archimedes and the Invention of Artillery and Gunpowder," *Technology and Culture* 28 (1987): 67–79. The claim also appears that Archimedes' fellow Syracusan Santa Lucia, martyred under Diocletian, was his descendant. On this story, see Simms, "Santa Lucia and Archimedes," *CA News* 13 (1995): 9–10.

9. It is one of "antagonistic and polemic" playfulness, which manifests itself in in-tricacy, elegance, and surprise. See Netz, *Works of Archimedes*, 11, 20, 25. In his forthcom-ing book *Ludic Proof: Greek Mathematics and the Alexandrian Aesthetic* (Cambridge: Cambridge University Press), Netz expands this argument, placing Archimedes' playful writing in the context of Hellenistic poetics. On epistolary prefaces as possibly an Archimedean innovation, see T. Janson, *Latin Prose Prefaces* (Stockholm: Almquvist and Wiksell, 1964), 19–23.

10. Reviel Netz, *The Shaping of Deduction in Greek Mathematics: A Study in Cogni-tive History.* (Cambridge: Cambridge University Press, 1999); see especially chap. 7, "The Historical Setting." See also Reviel Netz, "Greek Mathematicians: A Group Picture," in

Science and Mathematics in Ancient Greek Culture, ed. C. J. Tuplin and T. E. Rihll (Oxford and New York: Oxford University Press, 2002), 196–216.

11. J. L. Heiberg, ed., *Archimedes Opera Omnia cum Comentariis Eutocii* (Leipzig: Teubner, 1880–81; 2nd ed., 1910–15); on the palimpsest, see Heiberg, "Eine Neue Archimedeshandschrift," *Hermes* 42 (1907): 235–303; Netz, *Works of Archimedes*, 2. Netz and William Noel, the curator of manuscripts and rare books at the Walters Art Gallery, tell the riveting story of the palimpsest in *The Archimedes Codex* (London: Weidenfeld and Nicolson, 2007).

12. Netz, *Works of Archimedes*, 11. For Netz's fuller discussion of Archimedes' father, Pheidias, see Netz and Noel, *Archimedes Codex*, chap. 2. Netz also points out that Archimedes' name, read properly, in reverse, probably means "the Mind of the Principle" (as in the principle ordering the cosmos), instead of "the Number One Mind."

13. G. Libri, *Histoire des Sciences mathématiques en Italie, depuis la Renaissance des Lettres jusqu'à la fin du dix-septième siècle*, 4 vols. (Paris: J. Renouardi, 1838–41), 1:xi. Libri finished only four of the projected six volumes of this work, the first volume of which is taken up almost entirely by the introduction. Libri's life story is worth pursuing on its own account. A good starting point is the biography by P. Alessandra Maccioni Ruju and M. Mostart, *The Life and Times of Guglielmo Libri (1802–1869), Scientist, Patriot, Scholar, Journalist, and Thief: A Nineteenth-Century Story* (Hilversum: Verloren Publishers, 1995); on the *Histoire*, see 110–19.

14. Libri, *Histoire*, 1:xii.

15. Libri, *Histoire*, 1:3, 34–40. Ruju and Mostart (*Life and Times of Guglielmo Libri*, 116) note that Libri's history presents a view of Roman learning unusually critical for an Italian; however, T. J. Cornell (*The Beginnings of Rome: Italy and Rome from the Bronze Age to the Punic Wars (c. 1000–264 BC)* [London and New York: Routledge,1995], 151) points out that the discovery of the monuments and cemeteries of Etruria in the eighteenth century stirred interest in Etruscan Italy among Italian scholars and that this movement, having continued on into the nineteenth century, played a role in the Risorgimento. The Etruscans were fascinating because they were pre-Roman yet civilized and literate and because they were supposed to have united most of Italy. Cornell also points out a strong anti-Roman strain in the works of these Italians.

16. Libri, *Histoire*, 1:34: "La postérité leur reprochera à tout jamais la mort d'Archiméde."

17. Libri, *Histoire*, 1:35 n. 3. He quotes Cicero's words at *Tusc.* 5.64: *humilem homunculum a pulvere, et radio excitabo, . . . Archimedem.* On this passage, see chap. 2 in the present study.

18. Libri, *Histoire*, 1:37, with the quotation from Leonardo da Vinci on 208. See D. L. Simms, "Archimedes' Weapons of War and Leonardo," *British Journal for the History of Science* 21 (1988): 195–210.

19. Libri, *Histoire*, 1:38–39. He adds that although one can kindle wood from a distance with mirrors, there remain serious difficulties in setting fire to moving ships under combat conditions; moreover, Archimedes had other effective means of defending his city against an enemy fleet. Libri names no ancient sources, but the earliest surviving references to this story are the second-century CE Apuleius (*Apologia* 16) and Lucian (*Hippias* 2). On the mirrors, see Heiberg, *Quaestiones*, 39–41, with bibliography; D. L. Simms, "Archimedes and the Burning Mirrors at Syracuse," *Technology and Culture*

16 (1977): 1–18; W. R. Knorr, "The Geometry of Burning Mirrors in Antiquity," *Isis* 74 (1983): 53–73; Paul Keyser, "Archimedes and the Pseudo-Euclidean *Catoptrics:* Early Stages in the Ancient Geometric Theory of Mirrors," *Archives Internationales d'Histoire des Sciences* 35 (1985):114–15; (1986): 28–105. A 1992 paper by A. A. Mills and R. Clift argued that the ships could not be burnt with mirrors (see "Reflections on the 'Burning Mirrors of Archimedes' with a Consideration of the Geometry and Intensity of Sunlight Reflected from Plane Mirrors," *European Journal of Physics* 13 [1992]: 268–79), but recently a group of students from MIT has gone and done it, once again, under ideal conditions: see John Schwartz, "Recreating an Ancient Death Ray," *New York Times,* October 18, 2005.

20. Libri, *Histoire,* 37–38. The full quote reads: "Un fait qui mérite beaucoup plus d'attention, et qui a passé jusqu'à présent presque inaperçu, c'est qu'Archimède dut s'abaisser jusqu'à diriger la construction d'un vaisseau, où était une chambre destinée aux plaisirs honteux du roi. Voilà à quel prix il fut protégé par Hiéron!" For the expression in Athenaeus (ἑξῆς δὲ τούτων Ἀφροδίσιον κατεσκεύαστο τρίκλινιον, 5.207e), Libri (38 n. 1) offers up the following translation (with no attribution; it may be his own): *post haec ad Veneris voluptates aphrodisium extructum fuit, tribus lectis instructum.* The phrase *ad Veneris voluptates* (for the pleasures of Venus) appears to be a gloss of the Greek Ἀφροδίσιον.

21. At least this is the only whiff of sexuality until Gillian Bradshaw's fictional account of the romance between Archimedes and Hieron's half sister in *The Sand-Reckoner* (New York: Forge, 2000). There is not, alas, much sex to speak of here either. The fictional Delia and equally fictional young Archimedes play intricate duets on the auloi, even as our hero ponders how to construct siege engines and how to retain intellectual independence despite the lucrative defense contracts offered by Hieron.

22. I quote the translation of Ruju and Mostert in *Life and Times of Guglielmo Libri,* 119; see 151–56 on Libri's patriotism and advocacy of pure science.

23. Libri, *Histoire* 3: "Archimède éclipse les savans [*sic*] de la Grèce."

24. Libri, *Histoire,* 37. For claims to have discovered unedited sources, see Libri, *Histoire,* xxiv–xxv, on which see Ruju and Mostart, *Life and Times of Guglielmo Libri,* 115–16.

25. Sometimes, though, it does literally disappear. As Robert Gallagher has noted ("Metaphor in Cicero's *De Re Publica*," *CQ* n.s. 51, no. 2 [2001]: 509–19), J. Zetzel omits the passage on Archimedes' sphere in his edition (*Cicero, "De Re Publica," Selections* [Cambridge: Cambridge University Press, 1995]). See chap. 3 in the present study.

26. See chap. 2, on Cicero's discovery of Archimedes' tomb.

27. For a stimulating reading of the tradition, see Michael Authier, "Archimedes: The Scientist's Canon," in *A History of Scientific Thought: Elements of a History of Science,* ed. M. Serres (Oxford and Cambridge, MA: Blackwell, 1995), 124–59. Authier finds in the sources, especially Plutarch, the canonical "passion" narrative of the scientist sacrificed to political interests and only later rehabilitated. It does not fall within the scope of Authier's essay to explore why it is in the interests of authors besides Plutarch to shape such a narrative. Authier argues, I think rightly, that Plutarch's Platonism guides his presentation. See chap. 5 in the present study.

28. For a clear and useful survey of the history of ancient math, see S. Cuomo, *Ancient Mathematics* (London and New York: Routledge, 2001). In addition to works already noted, Simms's articles include "Galen on Archimedes: Burning Mirror or Burning Pitch?" *Technology and Culture* 32 (1991): 91–96. Simms also considers the reuse of

Archimedes' image in the modern world. See D. L. Simms and D. J. Bryden, "Archimedes as an Advertising Symbol," *Technology and Culture* 34 (1993): 387–92, in which the authors examine the use of Archimedes in the signs and handbills of London opticians from the seventeenth through the nineteenth centuries. A more complete list of Simms's contributions appears in the bibliography to the present study.

29. M. Clagett's essay "Biographical Accounts of Archimedes in the Middle Ages" (*Archimedes in the Middle Ages*, app. 3) has been invaluable for this project; so has Antonio Quacquarelli's "La fortuna di Archimede nei retorie negli autori cristiani antichi," *Rendiconti del Seminario Matematico di Messina* 1 (1960–61): 10–50, reprinted in *Saggi patristici—Retorica ed esegesi biblica*, Quaderni di *Vetera Christianorum* 5 (Bari: Adriatica Editrice, 1971), 381–424.

30. T. P. Wiseman, *Remus: A Roman Myth* (Cambridge and New York: Cambridge University Press, 1995); *The Myths of Rome* (Exeter: University of Exeter Press, 2004).

31. Judith Ginsburg, *Representing Agrippina: Constructions of Female Power in the Early Roman Empire* (Oxford and New York: Oxford University Press, 2006), 54.

32. Plut. *Marc.* 17.4; Vitr. *De arch.* 7 pr. 14. The text of Vitruvius's passage is problematic because the phrase referring to an anthology does not appear in the oldest manuscripts. Apuleius (*Apol.* 16.1–6) refers to a hefty volume on mirrors (*quae tractat uolumine ingenti Archimedes Syracusanus*), and Cassiodorus (*Var.* 1.45.4) reminds Boethius of having translated Archimedes' mechanical works into Latin. See Quacquarelli, "La fortuna di Archimede," 389–90.

33. Cic. *Luc.* 116 (= *Acad. Pr.* 2.11) may, however, suggest some awareness of the ideas behind *The Sand-Reckoner.* On Cicero's attitude toward and knowledge of mathematics, see E. Rawson, *Intellectual Life in the Late Roman Republic* (Baltimore: Johns Hopkins University Press, 1985),156–69. Paul Keyser ("Cicero on Optics (*Att.* 2.3.2)," *Phoenix* 47 [1993]: 67–69) is more optimistic as to Cicero's (or at least his architect's) having read Archimedes. He suggests that Cicero's use of ὄψις for eye is similar to that of Archimedes, who uses the same word for the pupil of the eye. Keyser points out that this is the only time Cicero wrote about optics, adding, "there is nothing unlikely in Cicero citing Archimedes—note the encomium at *Tusc.* 5.64–66" (67).

34. The seventeenth-century geometer Taquet wrote of Archimedes, "But more praise than would read, more wonder at than would understand him" (*Sed illum plures laudant quam legant; / admirantur plures quam intelligant*). This statement is quoted as an epigraph in P. Ver Eecke, *Les Oeuvres Complètes d'Archimède,* 1:vii. For the comparison to Einstein, see Netz and Noel, *Archimedes Codex,* chap. 2.

35. Tzetzes *Chil.* 2, 103–5; Polyb. 8.6.

36. For an overview of events in Sicily during the Second Punic War, see J. Briscoe, "The Second Punic War," *CAH,* 2nd ed. (1989): 8.44–80, especially 61–62; on Sicily as a province, see A. E. Astin, "Sources," *CAH,* 2nd ed. (1989): 8.1–16. See also R. J. A. Wilson, *Sicily under the Roman Empire: The Archaeology of a Roman Province, 36 BC–AD 535* (Warminster: Aris and Phillips, 1990), 18; J. Serrati, "The Coming of the Romans: Sicily from the Fourth to the First Centuries BC," in *Sicily from Aeneas to Augustus: New Approaches to Archaeology and History,* ed. C. Smith and J. Serrati. (Edinburgh: Edinburgh University Press, 2000), 109–14. For detailed analysis of the relationship between Rome and Sicily, see A. M. Eckstein, *Senate and General: Individual Decision-Making and Roman Foreign Relations, 264–194 B.C.* (Berkeley, Los Angeles, and London: University of California Press, 1987), 73–184.

37. E. Rawson, "Roman Tradition and the Greek World," *CAH*, 2nd ed. (1989): 8.428–30.

38. For a survey of these developments, see G. B. Conte, *Latin Literature: A History*, trans. J. B. Solodow, rev. D. Fowler and G. W. Most (Baltimore and London: Johns Hopkins University Press, 1994), 39–91; Rawson offers a more detailed overview in "Roman Tradition" (422–76). For current debate, see Feeney, "Beginnings of a Literature in Latin."

39. On Pictor, see H. Beck, "Den Ruhm nicht teilen wollen: Fabius Pictor und die Anfänge des römischen Nobilitätsdiskurses," in *Formen römischer Geschichtsschreibung von den Anfägen bis Livius*, ed. U. Eigler, U. Gotter, N. Luraghi, and U. Walter (Darmstadt: Wissenschaftliche Buchgesellschaft, 2003), 73–92, with recent bibliography. Beck stresses the coherence that Pictor's history produced by featuring the nobility.

40. See especially Erich S. Gruen, *Studies in Greek Culture and Roman Policy* (Leiden and New York: Brill, 1990); Erich S. Gruen, *Culture and National Identity in Republican Rome* (Ithaca: Cornell University Press, 1992); Thomas N. Habinek, *The Politics of Latin Literature* (Princeton: Princeton University Press, 1998). There has been a great deal of work done lately on Roman means of memorializing the past during the middle Republic. See, e.g., K.-J. Hölkeskamp, "*Exempla* und *mos maiorum*: Überlegungen zum kollektiven Gedächtnis der Nobilität," in *Vergangenheit un Lebenswelt: Soziale Kommunikation, Traditionsbildung und historisches Bewusstsein*, ed. H.-J. Gerhrke and A. Möller (Gunter Narr Verlag: Tübingen, 1996): 301–33; Harriet Flower, *Ancestor Masks and Aristocratic Power in Roman Culture* (Oxford: Oxford University Press, 1996). On the creativity of Roman writers using Greek models, see S. Hinds, *Allusion and Intertext: Dynamics of Appropriation in Roman Poetry* (Cambridge: Cambridge University Press, 1998), 52–63 (on Ennius and Naevius); Ingo Gildenhard, "The 'Annalist' before the Annalists: Ennius and his *Annales*," in Eigler et al., *Formen römischer Geschichtsschreibung*, 93–114; and, recently Feeney, "Beginnings of a Literature in Latin," 229, which emphasizes that the emergence of a literature in Latin should not be seen as inevitable.

41. For example, tradition credited Syracusans with the invention of systematic rhetoric (Cic. *Brut.* 46–48, reporting Aristotle's identification of Corax and Tisias as inventors of rhetoric in the early to middle fifth century), tradition reports that Plato's travels took him to Syracuse (Plato *Ep.* 7), and tradition identified the poet Theocritus as Syracusan. Theocritus set some of the *Idylls* in the surrounding countryside and addressed one (*Id.* 16) to Hieron II. On the early history of Syracuse, see T. J. Dunbabin, *The Western Greeks: The History of Sicily and South Italy from the Founding of the Greek Colonies to 480 B.C.* (Oxford: Clarendon, 1948). On its prosperity before the Second Punic War and its subsequent history under the Romans, see R. A. J. Wilson, *Sicily under the Roman Empire*, especially 17–32; Eckstein, *Senate and General*, 102–34, 156–84.

42. See, e.g., Th. Harrison, "Sicily in the Athenian Imagination: Thucydides and the Persian Wars," in Smith and Serrati, *Sicily from Aeneas to Augustus*, 84–96, with bibliography.

43. Thus Livy (21.1.1), invoking Thucydides (1.1–2), calls the Second Punic War the most memorable ever, and Silius Italicus models Saguntum on Plataea. On Livy, see B. S. Rogers, "Great Expeditions: Livy on Thucydides," *TAPA* 116 (1986): 335–52. On Silius, see W. Dominik, "Hannibal at the Gates: Programmatising Rome and *Romanitas* in Silius Italicus' *Punica* 1 and 2," in *Flavian Rome: Culture, Image, Text*, ed. A. J. Boyle and W. J. Dominik (Leiden and Boston: Brill, 2003), 469–97.

44. Thus the "Life of Archimedes" lends itself nicely to the study of what Matthew Roller has recently called the "discourse of exemplarity," a discourse involving four steps: an action; a primary audience, which judges the action; commemoration of the action, often by monuments; and imitation of the action by either the primary or a secondary audience. See M. Roller, "Exemplarity in Roman Culture: The Cases of Horatius Cocles and Cloelia," *CP* 99, (2004): 1–56.

45. In considering readers (both ancient and modern) and the act of reading and reception, I have found particularly useful Lowell Edmunds's discussion of reader-response theory and its application to classical literature. Edmunds explains: "In my view, no binarism of text and reader is possible and therefore no either-or decision has to be made. Meaning, including intertextual meaning, emerges from the interaction between the two" (*Intertextuality and the Reading of Roman Poetry* [Baltimore: Johns Hopkins University Press, 2001], 62). Although Edmunds is speaking of poetic intertextuality in poetry, his words apply to prose as well.

46. See especially Gerard Genette, *Narrative Discourse: An Essay in Method*, trans. J. E. Lewin (Ithaca: Cornell University Press, 1980); Mieke Bal, "Notes on Narrative Embedding," *Poetics Today* 2, no. 2 (1981): 41–59; Mieke Bal, "The Narrating and the Focalising: A Theory of the Agents in Narrative," *Style* 17 (1983): 234–69; Mieke Bal, "The Point of Narratology," *Poetics Today* 11, no. 4 (1990): 727–53.

47. *Histoire, récit,* and *narration* are Genette's terms to identify a set of events, the narrative of events ("the signifier, statement, discourse or narrative text itself"), and the act of narrating. See *Narrative Discourse*, 27.

48. Thus, for example, to refer to a narrator who is present as a character in his own story, I do not use Genette's term *homodiegetic narrator;* rather, I refer to "a narrator present as a character in his own story" or use words to that effect.

49. F. W. Walbank, *A Historical Commentary on Polybius*, 3 vols. (Oxford: Oxford University Press, 1957–77), 1:33. Polybius came to Rome as a hostage in 167 BCE and remained there until he accompanied Scipio Aemilianus to Spain in 151. It is generally agreed that the sixteen years that he spent in Rome stimulated him to begin his project. Some aged veterans of the Sicilian campaign (e.g., Cato) would have been around to interview.

50. Cic. *Clu.* 87; *Att.*, 12.15; 13.28; *Rep.* 1.21–22; *Acad. Pr. (Luc.* 116); *Fin.* 5.50; *Nat. D.* 2.88; *Tusc.* 1.63, 5.64–65.

51. Andrew Laird, *Powers of Expression, Expressions of Power: Speech Presentation and Latin Literature* (Oxford: Oxford University Press, 1999), xiii.

52. A highly informal survey of university students shows that this is the most familiar story about Archimedes. Most can associate Archimedes, the bath, and the cry "Eureka!" Fewer can tell what his discovery was. Sherman Stein gave his 1999 survey of Archimedes' mathematical contributions the title *Archimedes: What Did He Do Besides Cry Eureka?* (Washington, DC: Mathematical Association of America). He nicely summarizes the sources for Archimedes' life (1–6).

53. Behind Cicero (and beyond the scope of this study) stands the shadowy figure of Posidonius, Stoic philosopher and polymath, who made a planetary sphere (Cic. *Nat. D.* 2.88) and wrote about Marcellus's activities in Sicily (Plut. *Marc.* 20.7). On Posidonius, see Arnaldo Momigliano, *Alien Wisdom: The Limits of Hellenization* (Cambridge: Cambridge University Press, 1975), 22–49.

54. Habinek, *Politics of Latin Literature*, 34–68. For an earlier articulation of some of

Habinek's ideas, see his "Ideology for an Empire in the Prefaces to Cicero's Dialogues," *Ramus* 23, nos. 1–2 (1994): 55–67.

55. Laird, *Powers of Expression,* 57; see 44–78, especially 54–57 for discussion of *Rep.* 393d–394b1.

56. The sources are Cicero's *De republica* (54–51 BCE) and *Tusculan Disputations* (45 BCE) and Vitruvius's *De architectura* (between 27 and 23 BCE, although the dating is disputed: see B. Baldwin, "The Date, Identity, and Career of Vitruvius," *Latomus* 49 [1990]: 425–34).

CHAPTER 1

1. Hannibal took a core sample of a golden column at the temple of Juno at Croton (Cic. *Div.* 1.48) and, finding it to be solid gold, decided to steal the whole column. Juno appeared to him in a dream and said that if he did so, she would deprive him of his one good eye. In recompense, Hannibal made a golden calf from the sample he had taken and affixed it to the top of the column. See D. S. Levene, *Religion in Livy* (Leiden: Brill, 1993), 68. In comparison, because it makes Hieron rely on Archimedes to disclose the crime, rather than on the gods to take revenge, the crown story illustrates the king's confidence in Archimedes' intelligence and thus reinforces Archimedes' reputation for superhuman genius. Cf. Cic. *Rep.* 1.22: *plus in illo Siculo ingenii quam videretur natura humana ferre potuisse* ([that there was] more genius in that Sicilian than human nature seemed able to admit of). See discussion in chap. 3 of the present study.

2. Netz (Netz and Noel, *Archimedes Codex,* chap. 2) calls this an urban legend. He notes that the core of this story is the prescientific observation that "bigger things make bigger splashes." Netz adds that Vitruvius or his source (I think his source), probably was aware that Archimedes knew something about bodies immersed in water, but probably was not familiar with *On Floating Bodies,* the work that contains what we know as the Archimedes Principle. See also Netz, "Proof, Amazement, and the Unexpected," *Science* 298 (2002): 967–68.

3. Indeed, the *OED* (s.v. *eureka*) distinguishes between the word's original appearance in this significant context and its subsequent use: "The exclamation ('I have found it') uttered by Archimedes when he discovered the means of determining (by specific gravity) the proportion of base metal in Hiero's golden crown. . . . Hence *allusively,* an exulting exclamation at having made a discovery."

4. See, e.g., the circumspect entry for Archimedes in the third edition of the *OCD.* The sources, respectively, for *identidem* and πολλάκις are Vitr. *De arch.* 9 pr. 10 and Plut. *Mor.* 1093D.

5. Vitr. *De arch.* 9 pr. 9–12; Plut. *Mor.* 1093D (*Non posse suaviter vivi secundum Epicurum*).

6. Proclus 1.63, trans. Glen Morrow, in *Proclus, "A Commentary on the First Book of Euclid's 'Elements'"* (Princeton: Princeton University Press, 1970, 51. For Favinus's poem, see *Anth. Lat.* 1.485, in F. Buechler and A. Riese, eds. *Anthologia Latina* (Leipzig: Teubner, 1906; reprint, Amsterdam: Hakkert, 1964) vol. 1.2.29–37.

7. J. Soubiran, trans. *Vitruve, De l'architecture, Livre IX* (Paris: Les Belles Lettres, 1969), xxv. References to detecting the theft (*reprehenderet furtum; reprehendit*) open and close Archimedes' part of the story. Marcel Achard (*Sophocle et Archimède: Pères du roman policier* [Liège: Éditions Dynamo, 1960], 5) sees Archimedes and Sophocles as the

founding fathers of the *roman policier,* with Sophocles being the ancestor of such writers as Balzac, Dickens, Hugo, and Dumas and with Archimedes being the ancestor of the detectives in Poe, Voltaire, Conan Doyle, and Agatha Christie.

8. Indeed, astronomy, the relative motions of the planets, and the celestial spheres are the topic of the first part of the book (9.1.1–9.4.6).

9. On the commonplace and on the rhetoric of the preface, see Soubiran, *Vitruve,* xi, xvii–xxxiv. Jean-Marie André ("La Rhétorique dans les Préfaces de Vitruve: Le Statut Culturel de la Science," in *Filologia e Forme Letterarie: Studi Offerti a Francesco Della Corte,* ed. Sandro Boldrini [Urbino: Quattro Venti, 1987], 3:265–89) places Vitruvius's prefaces in the context of Ciceronian rhetorical theory. L. Callebat ("Rhétorique et architectura dans le De Architectura de Vitruve," in *Le Projet du Vitruve: Objet, Destinataires et Réception du De Architectura,* ed. P. Gros [Rome: L'École Française, 1994], 33) categorizes the prefaces as "suasiones ad architecturam," examples of deliberative rhetoric following the pattern of Cato's *De agricultura.*

10. By "Caesar," Vitruvius means Augustus, to whom the entire work is dedicated. On the name, see A. Fleury, ed. and trans., *Vitruve,* vol. 1 (Paris: Les Belles Lettres, 1990), 51–52.

11. On the date of composition of the *De architectura,* see Soubiran, *Vitruve,* xxxii; on *De arch.* 9 pr. 17, see Baldwin, "Date, Identity, and Career of Vitruvius."

12. Soubiran (*Vitruve,* xxxi) observes the symmetry.

13. Vitr. *De arch.* 9 pr. 1.

14. Cf. Vitr. *De arch.* 9 pr. 2: *Quid enim Milo Crotoniates, quod fit invictus, prodest hominibus aut ceteri, qui eo genere fuerunt victores, nisi quod, dum vixerunt ipsi, inter suos cives habuerunt nobilitatem?* (For what good did Milo of Croton do men in being undefeated, or others who were victors of that kind, except that, while they themselves lived, they had distinction among their own fellow citizens?).

15. The contributions of writers unite the *utile* and *honestum.* André ("La Rhétorique," 287–88) sees in Vitruvius's idea of a unifying culture the influence of Cicero's *De oratore* and *Pro Archia.*

16. On Isocrates, see *Paneg.* 1–2, Soubiran, *Vitruve,* xxi. I thank Andrew Feldherr for drawing my attention to the significance of this reference. Vitruvius uses *scriptores* three times in *De arch.* 9. pr. 1–3 and refers to the benefits of books in *De arch.* 9 pr. 1.

17. Plutarch's version (quoted later in the present chapter) neither mentions Archimedes' nakedness nor says that he ran. Plutarch probably takes it for granted that his audience (readers of Greek, during the early empire, in a culture accustomed to public baths) will assume the nakedness of a bather, leaving no need to mention it. On the increased attention to baths during the early empire, see F. Yegül, *Baths and Bathing in Classical Antiquity* (Cambridge, MA: MIT Press, 1992): 30–47, especially 40. Christopher Hallett (*The Roman Nude: Heroic Portrait Statuary, 200 BC–AD 300* [Oxford: Oxford University Press, 2005], 62) points out that literary sources do not refer to complete nudity every time they use the word *nudus,* for it most commonly meant "lightly clad" rather than completely naked. In fact, says Hallett, "*nudus* can only be taken to mean fully naked in circumstances where complete nudity is to be expected—as for example in the Greek gymnasium, the stadium, or the baths"; that is, *nudus* means "bare-naked" in circumstances in which the word is slightly redundant. For full discussion, see Hallett, *Roman Nude,* 61–101.

18. Yegül, *Baths and Bathing,* 24; see 23 on the fusion of gymnasia and baths as a

phenomenon of the Hellenistic period, 49 for a good photo of some Sicilian (Gela) hip baths from this period.

19. Hallett (*Roman Nude*, 21–23, 25–27) points out that the tradition of Hellenistic portraiture included nude statues of victorious athletes, as well as funerary stelae depicting nude youths with the oil flask that was part of the athlete's equipment. He also observes (80) that the walls of the palestrae of the baths were often decorated with depictions of nude Greek athletes, images that did not reflect the real state of Roman undress but were "a kind of homage to the Greek athletic ideal."

20. Romans of Vitruvius's time could already have been familiar with the running nude as iconic for the Greek athlete. A silver denarius minted at Rome in 74 BCE depicts an athlete, naked and running, with boxing gloves in one hand and the palm of victory in the other. See Hallett, *Roman Nude*, 68–69, plate 35 (= M. H. Crawford, *Roman Republican Coinage* (Cambridge: Cambridge University Press, 1974), RRC 396/1). Indeed, the very object central to the story, Hieron's *corona votiva*, picks up the idea of the crown (*corona*) awarded the athletic victor. The eighteenth-century engraving on the title page of Mazzuchelli's *Notizie* shows the naked Archimedes carrying off the crown as he runs. Mazzuchelli's image rather resembles that of the athlete on the denarius carrying off the palm and the boxing gloves.

21. Paul Zanker (*The Mask of Socrates: The Image of the Intellectual in Antiquity*, trans. A. Shapiro [Berkeley and Los Angeles: University of California Press, 1995], 20–22) points out that in a sanctuary, such as at Olympia, a portrait of an elderly and decrepit Homer would have stood not far from statues of physically imposing youths (he gives the Riace Bronzes as examples): "The juxtaposition of two such antithetical statues must have intensified the contrast, the unbounded glorification of youth and strength in a warrior or athlete set against the wisdom of age and corporeal frailty."

22. Plutarch, *Moralia* 1094c; Trans. B. Einarson and P. H. De Lacy, in *Plutarch, "Moralia,"* vol. 14 (Cambridge, MA: Harvard University Press, 1967), 67–68.

23. *On Floating Bodies*, proposition 7. This is the Archimedes Principle (see Netz, "Proof, Amazement, and the Unexpected.")

24. E.g., Favinus *Carm.* 135: *quod te quale siet, paucis (adverte) docebo*; 138: *imponas*; 139: *summittas*; 142: *siste*; 145: *fac*; 150 *specta*.

25. Differences include the following: Vitruvius says that Archimedes leaped from his bath moved by joy, whereas Plutarch, by referring to his drawing on his belly, suggests a constant state of intellectual pleasure; Vitruvius and Favinus introduce the crown story by way of Hieron's request, while Proclus records the reaction of Hieron's son, Gelon, to Archimedes' discovery; Plutarch alone makes no mention of the royal family in the "Eureka" story (although he does when he discusses Archimedes in the *Marcellus*); Proclus's version preserves Hieron's words and Gelon's echo of them instead of Archimedes' cry; Vitruvius and Proclus note that the crown was an offering to the gods; Favinus and Proclus state that Archimedes solved the problem without damaging the offering—"without harming the item that had been already dedicated to the gods" (*inlaeso quod dis erat ante dicatum*) and "without destroying the crown that had been made" (τοῦ στεφάνου μὴ λυθέντος). This last detail seems to reflect a more problem-oriented approach to the story, as if the versions known to these fifth-century authors began with a set of givens and restrictions: given an irregularly shaped metal object, which one cannot damage, how does one find out its composition?

26. A review of the 131 uses of the first-person perfect in the *Thesaurus Linguae*

Graecae database shows that εὕρηκα is the standard expression for finding something. None repeats the word as Vitruvius and Plutarch do, nor, as far as I can tell, does any of the other uses parody Archimedes' expression.

27. To paraphrase Stephen Hinds (*Allusion and Intertext*, 1), Plutarch draws attention to the fact that he is alluding in this story, and he reflects on the nature of the allusive activity. Plutarch might be alluding specifically to Vitruvius, but it is more likely that he is gesturing toward a tradition, both oral and written. While Plutarch and Vitruvius are probably not engaging in allusive activity at the level of specificity with which Ovid, for example, would allude to an already allusive passage in Catullus, the invocation of tradition was part of the trained writer's rhetorical tool kit, whether that writer was working in poetry or prose. On Plutarch's use of allusion, see C. Pelling, *Plutarch and History: Eighteen Studies* (London: Classical Press of Wales; London: Duckworth, 2002), 122–24 (on Plutarch's use of Thucydides), 171–74 (on allusion in the *Life of Theseus*), 197–206 (on the role of Dionysius in the *Life of Antony* and the *Life of Theseus*), 275–76. Pelling points out: "Literary culture is also assumed, enough to welcome the quotations and allusion which lace his [Plutarch's] narrative; enough, even to catch allusions which the narrator does not label, confident that the narratee will be able to fill in the gap" (275).

28. A partial, not-surprisingly gold-oriented list of places and things whose names are derived from εὕρηκα includes Eureka, California; Eureka, Kansas; Eureka County, Nevada; Eureka Springs, Arkansas; a hotel in an Australian gold town that was the center of the Eureka Stockade uprising in 1854; Eureka College, Ronald Reagan's alma mater; a make of vacuum cleaner; the journal of the Mathematics Society of the University of Cambridge; and a recycling center in Minneapolis–St. Paul.

29. An exception is Henry James, who writes, "But in fine—with which I cry eureka, eureka; I have found what I want for Rosanna's connection" (*The Ivory Tower, Working Notes*, in *The Complete Notebooks of Henry James*, Leon Edel and Lyall H. Powers, eds. (Oxford: Oxford University Press, 1987), 471.

30. Pl. *Tht.* 174b; *Symp.* 174d–175a. See Netz, *Shaping of Deduction*, 304. Augustan accounts of exemplary Romans subordinating body to mind are stories not of absent-mindedness but of patriotic will. Such Romans make spectacles of themselves but in quite another way, as did, for example, Livy's Mucius Scaevola, who held his hand to the fire before the eyes of Lars Porsenna and his men (Livy 2.12.13). See Andrew Feldherr, *Spectacle and Society in Livy's History* (Berkeley and Los Angeles: University of California Press, 1998), 120; Roller, "Exemplarity" 1–4.

31. "Intellectual" is Zanker's shorthand term for poets, philosophers, thinkers, and orators; we do not have ancient images of Archimedes, but Zanker includes Plato, Pythagoras, and Eudoxus in his discussion. See Zanker, *Mask of Socrates*, 2; for discussion and pictures of third-century images of men thinking hard, see 90–145.

32. Hallett (*Roman Nude*, 61–68), discussing the Romans' taboo against being seen naked in public, points out that Romans thought of being stripped and exposed to public view as degradation to be inflicted on condemned criminals or as the unenviable fate of slaves displayed for purchase. During the first century BCE, nudity, even in the baths, was the exception at Rome rather than the rule. Literary sources suggest that one wore some kind of garment, unless one was actually bathing or being rubbed down. Hallett (82) also notes that casual references to complete nudity in the baths are absent from the literature of the Augustan Age. See also Yegül, *Baths and Bathing*, 30–47, especially 34.

Hallett (*Roman Nude*, 68–69) notes that according to Dionysius of Halicarnassus, the Romans of the late first century BCE, never exercised fully naked, as the Greeks did. For images of gladiators and athletes from a later period (Baths of Caracalla), see Yegül, *Baths and Bathing*, 161, fig. 177.

33. C. Edwards, *The Politics of Immorality in Ancient Rome* (Cambridge and New York: Cambridge University Press, 1993), 63–97; A. Corbeill, *Nature Embodied: Gesture in Ancient Rome* (Princeton and Oxford: Princeton University Press, 2004), 117–24. See also Karl Galinsky, *Augustan Culture: An Interpretive Introduction* (Princeton: Princeton University Press, 1996), 14. On self-control in classical Athens, see Zanker, *Mask of Socrates*, 48–49. To elite readers, Archimedes' disregard of self-presentation would have placed him on the margins of society.

34. Corbeill, *Nature Embodied*, 69.

35. Corbeill, *Nature Embodied*, 69, 128.

36. His image calls to mind the following address from Martial to Afer (Mart. 6.77.5–6), written a full century after Vitruvius: *rideris multoque magis traduceris, Afer, / quam nudus medio spatiere foro* (you are laughed at and exposed to scorn much more, Afer, than if you were to stroll naked in the middle of the forum). The passage is quoted by Hallett (*Roman Nude*, 83 and n. 43), who notes the persistence of the Roman taboo against being seen naked in public.

37. On public scrutiny and aristocratic honor, see J. E. Lendon, *Empire of Honour: The Art of Government in the Roman World* (Oxford: Oxford University Press, 1997), 36–61.

38. Cic. *Phil.* 2.65.8: *in eius igitur viri copias cum se subito ingurgitasset, exsultabat gaudio persona de mimo, modo egens, repente dives.* See J. T. Ramsey, ed., *Cicero, "Philippics" I–II* (Cambridge: Cambridge University Press, 2003). Ramsey (255) points out that *Suddenly Rags to Riches* (*modo egens, repente dives*) might have been the title of a play, although it is attested nowhere else.

39. Corbeill (*Nature Embodied*, 117) observes: "The audience of Roman comedies was also expected to recognize correlations between movement and character—members of the dominant class move slowly upon the stage, whereas slaves, attendants, and workers were marked by sterotypically swift movements." He also notes that "the 'running slave' appears so often in Roman comedy as to render the expression almost tautological." On the figure of the running slave, see, e.g., G. E. Duckworth, *The Nature of Roman Comedy: A Study in Popular Entertainment,* 2nd ed. (Norman: University of Oklahoma Press, 1994), 106–7; J. Wright, *Dancing in Chains: The Stylistic Unity of the Comoedia Palliata,* Papers and Monographs of the American Academy in Rome 25 (Rome: American Academy in Rome, 1974), 114–15 and passim. For the appearance of this stock character in noncomedic contexts, see J. A. Harrill, "The Dramatic Function of the Running Slave Rhoda (Acts 12.13–16): A Piece of Greco-Roman Comedy," *New Testament Studies* 46 (2000): 150–57. Harrill interprets the slave's entrance as that of a *serva currens* and views it as Luke's attempt to provide comic relief.

40. The figures of running slave and running athlete overlap explicitly in Plautus's *Stichus,* where the running slave Pinacium compares himself to a competitor training for the Olympic Games (306): *simulque cursuram meditabor ad ludos Olympios* (likewise I'll practice my running for the Olympic games). Even Vitruvius's description of the triumphs deserved by great writers (*De arch.* 9 pr. 2) can be read in a comic context, for the clever slave (*servus callidus*) uses the triumph metaphor when describing his own exploits. See also Plaut. *Asin.* 269; *Bacch.* 1068–75. See Wright, *Dancing in Chains,* 105–6.

Wright also points out that one of Caecilius Statius's plays, which probably featured a *callidus servus,* was titled *Triumphus.*

41. Netz, *Shaping of Deduction,* 279–280; Netz, *Works of Archimedes,* 11; Netz and Noel, *Archimedes Codex,* chap. 2.

42. Sil. It. *Pun.* 14.343; Plut. *Marc.* 14.6–14. To use terms developed by Pierre Bourdieu (*Distinction: A Social Critique of the Judgement of Taste,* trans. R. Nice [Cambridge, MA: Harvard University Press, 1984]; *The Field of Cultural Production: Essays on Art and Literature,* ed. R. Johnson [New York: Columbia University Press, 1993]), Cicero and Silius deprive Archimedes of economic capital (material resources) and social capital (in their stories, he does not have connections to the ruling elite). For Cicero, this is a way of placing emphasis on Archimedes as a self-made creator and owner of cultural capital, like Cicero himself (see chap. 3 in the present study); for Vitruvius, as we shall see, Archimedes is literally an owner of "embodied" capital.

43. The entire work is dedicated to Augustus. The opening of the preface to book 1, it has been noted, bears a striking resemblance to the poet Horace's address to Augustus in *Ep.* 2.1.1–4. See Baldwin, "Date, Identity, and Career of Vitruvius."

44. Vilr. 9 pr. 9; Augustus *Res Gestae* 34.3.

45. On Vitruvius's status anxiety, his concern for the *dignitas* of the architectural profession, and the connection of *dignitas* to the *auctoritas* of buildings, see M. Masterson, "Status, Pay, and Pleasure in the *De Architectura* of Vitruvius," *AJP* 125 (2004): 387–416. Masterson shows how Vitruvius uses an anecdote about Aristippus in the preface to book 6 as a way of exploring issues of status, pay, and pleasure; by doing so, he locates Vitruvius's anxiety about his status as an *apparitor* within the social context of the Augustan cultural revolution.

46. In the preface to book 2 of the *De architectura,* Vitruvius presents himself as a writer who is particularly unimpressive in person, saying, "To me, however, Imperator, nature did not give height. Age has ruined my appearance, and ill health has drawn away my strength. Therefore, abandoned by these protections, I shall find my way to your favorable notice using the auxiliary forces of knowledge and my writing" (*Mihi autem, imperator, staturam non tribuit natura, faciem deformavit aetas, valetudo detraxit vires. Itaque quoniam ab his praesidiis sum desertus, per auxilia scientiae scriptaque, ut spero, perveniam ad commendationem,* 2 pr. 4). Vitruvius and Archimedes may share some physical features. Although no source reports Archimedes' age at the time of his famous bath, Zanker (*Mask of Socrates,* 22) observes that "in the Greek imagination, all great intellectuals were old." Cicero's calling Archimedes a *homunculus* (little man) at *Tusc.* 5.64 may have more to do with social status than physical size. Yet age and lack of gravitas would make the mathematican, like the architect, a physically unprepossessing intellectual. Writing is a way of turning age, at least, to advantage, according to Indra McEwen (*Vitruvius: Writing the Body of Architecture* [Cambridge, MA: MIT Press, 2003], 103): "Writing, Vitruvius claims, gives an architect *auctoritas,* that specficially Roman measure of worth. . . . How much more authoritative, then, an architect/author who is old. Or better still, dead." What characterizes Vitruvius's exemplary *scriptores* is their absence. Almost all—if not all—the authors that Vitruvius lists are dead (Varro may be the sole exception) and represented by their writing, which of course lives on, unlike their bodies and those of athletes. "Books," says McEwen, "are the missing bodies of writers consecrated by old age."

47. Forms of the verb *invenio* and the noun *inventio* appear sixteen times in this

preface (cf. twice in the preface to book 2, three times in each of books 7 and 8, and no occurrences elsewhere).

48. On this order, discovery then proof, see M. Clagett, *Greek Science in Antiquity* (New York: Abelard and Shuman, 1955; reprint, Minola, NY: Dover, 2001), 61–63. Vitruvius, of course, describes discovery by mechanical methods followed by mechanical proof, instead of the discovery by mechanical methods followed by geometrical proof that Archimedes describes in the introduction to his *Method.*

49. Vitruvius notes, to be sure, that Pythagoras sacrificed an ox to the Muses, because he felt that they had inspired his discovery, but he mentions this only after showing how the theorem works, and he presents it as proof of Pythagoras's gratitude, not his joy (*maximas gratias agens hostias dicitur his immolavisse, De arch.* 9 pr. 7).

50. Indra McEwen's *Vitruvius: Writing the Body of Architecture* takes as its starting point the metaphors of the body that govern the writing of the treatise. McEwen points out that "Vitruvius claims, repeatedly, that he was 'writing the body of architecture'" (6). Recognizing what she calls "the opacity of this metaphor" (5), McEwen argues that it bears a special relationship to Vitruvius's project and, indeed, is fundamental to it. McEwen's work draws on much of the recent Continental work on Vitruvius (e.g., on the connection between writing and architecture, see especially Callebat, "Rhétorique et architecture," 31–46). It will take some time for students of Augustan literature to unravel McEwen's rich and complicated web of ideas. For a critical yet sympathetic review of McEwen's *Vitruvius,* see R. Taylor's review in *CP* 100 (2005): 284–89; also helpful is that of L. A. Riccardi in *Aestimatio* 2 (2005): 136–41.

51. Vitr. *De arch.* 3.1.1–9.

52. *Quantum corporis sui in eo insideret, tantum aquae extra solium effluere* (Vitr. *De arch.* 9 pr. 10). Vitruvius uses the same phrasing to describe the bodies of athletes (*sua corpora,* 9 pr. 1; *suis corporibus,* 9 pr. 15).

53. Vitr. *De arch.* 9 pr. 10: "after the allegation was made that gold had been removed and the same quantity of silver had been blended in for that crown-making job . . ." (*posteaquam indicium est factum dempto auro tantundem argenti in id coronarium opus admixtum esse . . .*).

54. Leslie Kurke, *Coins, Bodies, Games, and Gold: The Politics of Meaning in Archaic Greece* (Princeton: Princeton University Press, 1999), 42.

55. Kurke, *Coins, Bodies, Games, and Gold,* 43.

56. Kurke, *Coins, Bodies, Games, and Gold,* 51–53.

57. Theognis 77–78, 415–18. See Kurke, *Coins, Bodies, Games, and Gold,* 44, 50.

58. Kurke, *Coins, Bodies, Games, and Gold,* 50.

CHAPTER 2

A version of this chapter appeared as "Cicero and Archimedes' Tomb," *JRS* 92 (2002): 49–61. I am grateful to the Roman Society for permission to reuse this article.

1. For images of this painting and others on the topic, see the plates accompanying J. B. Trapp's "Archimedes' Tomb and the Artists: A Postscript," *Journal of the Warburg and Courtauld Institutes* 53 (1990): 286–88. Two of Benjamin West's depictions of this scene appear in Helmut von Erffa and Allen Staley's *The Paintings of Benjamin West* (New Haven: Yale University Press, 1986), 120 (cat. no. 22, a version painted in 1797, now

in private hands) and 175 (cat. no. 23, the 1804 Yale picture). See also the images collected by Chris Rorres at http://www.mcs.drexel.edu/~crorres/Archimedes/contents.html.

2. In "Archimedes' Tomb," Trapp puts these paintings in the context of the late-eighteenth century vogue for discovering classical tombs. In biographies of Cicero, the tomb story usually follows a reference to Cicero's popularity in Sicily and preceeds a reference to the anecdote in which Cicero tells how he learned, on his return from Sicily, that no one in Rome knew or cared where he had been (*Planc.* 64). On Cicero, see, e.g., M. Fuhrmann, *Cicero and the Roman Republic*, trans. W. E. Yuill (Oxford and Cambridge, MA: Blackwell, 1992), 37; M. Gelzer, *Cicero: Ein Biographischer Versuch* (Wiesbaden: F. Steiner, 1969), 29, 308–9; C. Habicht, *Cicero the Politician* (Baltimore: Johns Hopkins University Press, 1990), 22; E. Rawson, *Cicero: A Portrait* (Ithaca: Cornell University Press, 1975), 33–34; E. G. Sihler, *Cicero of Arpinum: A Political and Literary Biography* (New Haven: Yale University Press, 1914), 64–65. On Archimedes, see, e.g., Heath, *Works of Archimedes*, 1–5; Dijksterhuis, *Archimedes*, 9–32. Exceptions to the rule are D. L. Simms ("The Trail for Archimedes's Tomb," *Journal of the Warburg and Courtauld Institutes* 53 [1990]: 281–86), who is interested in the tomb's physical appearance, and Serafina Cuomo (*Ancient Mathematics*, 197–98), whose remarks on the tomb appeared when the article on which this chapter was originally based was in press; parts of my interpretation agree in general outline with hers.

3. See especially A. Brinton, "Cicero's Use of Historical Examples in Moral Argument," *Philosophy and Rhetoric* 21, no. 3 (1988): 169–84; Thomas N. Habinek, "Science and Tradition in *Aeneid* 6," *HSCP* 92 (1989): 223–55. For a catalog of Cicero's exempla, see M. N. Blincoe, "The Use of the Exemplum in Cicero's Philosophical Works" (PhD diss., St. Louis University, 1941). Oddly enough, Blincoe does not refer to the Archimedes story. On historical exempla and their social function, see Martin Bloomer, *Valerius Maximus and the Rhetoric of the New Nobility* (Chapel Hill: University of North Carolina Press, 1992), especially 4–10 (on Cicero).

4. A. E. Douglas characterizes the *Tusculan Disputations* as "rhetorized" philosophy rather than philosophical rhetoric. See "Form and Content in the *Tusculan Disputations*," in *Cicero the Philosopher: Twelve Papers*, ed. J. G. F. Powell (Oxford: Oxford University Press, 1995), 197–218; see especially 198–200, with reference to *Tusc.* 1.7. See also P. MacKendrick and K. L. Singh, *The Philosophical Books of Cicero* (New York: St. Martin's Press, 1989), 165. Ann Vasaly (*Representations: Images of the World in Ciceronian Oratory* [Berkeley and Los Angeles: University of California Press, 1993], 257) puts it well when she concludes that Cicero's reliance on the concrete was "the Roman gateway to the world of ideas." A. D. Leeman (*Orationis Ratio: The Stylistic Theories and Practice of the Roman Orators, Historians, and Philosophers* [Amsterdam: Hakkert, 1963], 206) also stresses the personal: "Just as Aristotle had combined his philosophy with rhetorical studies, Cicero wants to add philosophy to his oratory. Of course he could not ignore the age-old feud between rhetoric and philosophy. But ever since his *De Oratore*—as we have seen—he had tried to bring the enemies together. Rhetoric and philosophy are indeed two different things, he makes Crassus argue, but they should be combined in *one and the same person*" (emphasis added).

5. Cicero (*Tusc.* 5.1) argues: *Nihil est enim omnium quae in philosophia tractantur quod gravius magnificentiusque dicatur* (For of all the topics treated of in philosophy, there is none to be spoken of with more dignity and splendor). Note that in justifying

his attempt to make this proof, Cicero also describes how it can be treated rhetorically. On the topic's position within the dialogue, see Douglas, "Form and Content in the *Tusculan Disputations*," 208–9.

6. *Tusc.* 5.55–62. The syllogistic argument, in 5.15–20, is nicely outlined in T. W. Dougan and R. M. Henry, eds., *Ciceronis Tusculanae Disputationes*, vol. 2, *Books III–V* (Cambridge: Cambridge University Press, 1905), xxii–xxiii: the man who is under the influence of emotions, such as fear, is unhappy; men subject to none are happy; this tranquillity is produced by virtue; therefore virtue suffices for living happily. For a more detailed outline of the argument in book 5, see H. A. K. Hunt, *The Humanism of Cicero* (Carlton: Melbourne University Press, 1954), 116–24.

7. Cicero uses the same pair at *Rep.* 1.28: *Quis enim putare vere potest, plus egisse Dionysius tum cum omnia moliendo eripuerit civibus suis libertatem, quam eius civem Archimedem cum istam ipsam sphaeram, nihil cum agere videretur, de qua modo dicabatur, effecerit?* (For who, truly, can suppose that Dionysius achieved more, when by all manner of contriving he tore their freedom away from his fellow citizens, than did his compatriot Archimedes, when, while he appeared to be achieving nothing, he made that very sphere that was just recently the topic of discussion?). Writing of Seneca, Roland Mayer ("Roman Historical Exempla in Seneca," in *Sénèque et la prose latine*, ed. O. Reverdin and B. Grange, Entretiens Hardt 36 [Geneva: Fondation Hardt, 1991], 141–69) observes that two devices helped to impose some control on otherwise shapeless lists of exempla: a tendency to group exempla into threes and the rhetorical crescendo, which determines the order of exempla within the list.

8. I follow here the text of Dougan and Henry, *Ciceronis Tusculanae Disputationes*.

9. For a list of Cicero's philosophical works, see Powell, *Cicero the Philosopher*, xiii–xvii. On the preface, see Douglas, "Form and Content in the *Tusculan Disputations*," 207–9. For an introduction to Cicero's philosophical works, see Douglas, "Cicero the Philosopher," in *Cicero*, ed. T. A. Dorey (New York: Basic Books, 1965), 135–70; Powell, *Cicero the Philosopher*, 1–35. On Cicero's politics, see Habicht, *Cicero the Politician*; D. Stockton, *Cicero: A Political Biography* (Oxford: Oxford University Press, 1971), 254–79; N. Wood, *Cicero's Social and Political Thought* (Berkeley and Los Angeles: University of California Press, 1988).

10. See Douglas, "Form and Content in the *Tusculan Disputations*," 206; Habinek, "Ideology for an Empire," 58–59. Habinek (*Politics of Latin Literature*, 64) observes, "Cicero makes it clear that what he is doing is not in fact Hellenizing his own practice, but rather Romanizing Greece." Habinek points out the metaphors of illumination (*inlustrandum*) and guardianship (*tuemur*) in the opening passage, both of which, as we shall see, are relevant to Cicero's response to the tomb. On the appropriation of Greek learning within Roman cultural norms, see P. L. Schmidt, "Cicero's Place in Roman Philosophy: A Study of His Prefaces," *CJ* 74 (1978–79): 115–27. Schmidt (119) observes that in addition to imparting philosophical instruction, the literary form of the dialogue continued sociocultural contacts by way of dedication and, in some cases, even served as an obituary notice or *laudatio funebris*.

11. I use here a metaphor developed by Susan Treggiari in "Home and Forum: Cicero between 'Public' and 'Private'" *TAPA* 128 (1998): 1–23. Treggiari argues against compartmentalizing Cicero's life and work into discrete categories of public, private, philosopher, orator, and statesman: "The individual's experience is a continuum. Let us think of this in relation to physical space, indoors and outdoors. In Roman thinking

about the house, public and private join up and overlap. The threshold of the house does not mark a barrier between public and private worlds, but a marker over which household members and non-members pass to go in or out" (3).

12. On *digressio*, see Cic. *Inv. rhet.* 1.27, 51, 97; Cic. *De or.* 2.80, 2.311–12, 3.203; Cic. *Brut.* 292, 322; J. C. Davies, "*Reditus ad Rem:* Observations on Cicero's use of *Digressio*," *Rheinisches Museum* 131 (1988): 305–15; H. V. Canter, "*Digressio* in the Orations of Cicero," *AJP* 52 (1931): 351–61. The *Pro Archia* comes to mind again here, since, as Canter (361) points out, the praise of Archias (*Arch.* 12–32) is one long digression.

13. MacKendrick (MacKendrick and Singh, *Philosophical Books of Cicero,* 25) points out: "The form of the Dialogues, as we have seen, is at least some of the time patterned on the traditional parts of a forensic speech. The recessed panel or ring-composition (A-B-C-D-C'-B'-A'), ideas patterned around a core, can be detected in the praise of philosophy (*Tusc.* 5.5). This is the kind of self-imposed restraint classical art thrives on."

14. A. E. Douglas calls it a "pleasing, but completely irrelevant, anecdote" (*Cicero, "Tusculan Disputations" II & V, with a Summary of III & IV* [Warminster: Aris and Phillips, 1990], note to 5.64). The apparent irrelevance marks this passage as a digression, thus helping it fulfill its rhetorical function.

15. On the structure of the argument, see Douglas, "Form and Content in the *Tusculan Disputations*," 197–204.

16. This is how later artists represent Archimedes' death. See, e.g., the charcoal by Honoré Daumier in B. Laughton, *The Drawings of Daumier and Millet* (New Haven: Yale University Press, 1991), 51.

17. The Roman commander Marcellus is said to have been both angered and aggrieved by the scientist's death (*permoleste tulisse*, Cic. *Verr.* 4.131). See also Livy 25.31.10; Plut. *Marc.* 19. I discuss these passages in chap. 4 of the present study.

18. In his dialogues, Cicero uses *indagare* and the noun *indagatio* for inquiry and the act of inquiry (*Luc.* 127; *Fin.* 5.58; *Off.* 1.15). At *Tusc.* 5.5, he addresses philosophy with *O virtutis indagatrix expultrixque vitiorum!* See Dougan and Henry, *Ciceronis Tusculanae Disputationes,* ad loc., on *indagatrix.* E. Fantham (*Comparative Studies in Republican Latin Imagery* [Toronto: University of Toronto Press, 1972], 146–47) gives other examples of Cicero's personifications.

19. On the effect of the word order, see O. Heine and M. Pohlenz, eds., *Ciceronis Tusculanorum Disputationum, Libri V,* vol. 2 (Stuttgart: Teubner, 1957), 131.

20. Cic. *Tusc.* 4.44: *Noctu ambulabat in publico Themistocles cum somnum capere non posset, quaerentibus respondebat Miltiadis tropaeis se e somno suscitari.* On this kind of competition, see also Cic. *Arch.* 24; Suet. *Iul.* 7.1; Sall. *Iug.* 4.5–6; *Anth. Lat.* 708 (an epigram addressed by Germanicus to Hector's tomb); Livy 28.43.6 (Scipio Africanus's words).

21. Plut. *Marc.* 17. The ratio could also be that of their surface areas. Both are 2:3. For more discussion of the Plutarch passage, see chap. 5 in the present study.

22. See Elizabeth P. McGowan, "Tomb Marker and Turning Post: Funerary Columns in the Archaic Period," *AJA* 99, no. 4 (1995): 615–32. According to McGowan, columns are attested as grave markers in the Greek world, including Sicily, as early as the Archaic Age. A sculpture on top of such a column might have particular relevance to the life of the person it commemorated: the column of Diogenes the Cynic, for example, was said to support a sculpture of a dog, a symbol of his philosophical school. Or such a sculpture might simply symbolize death, as does the figure of a Siren on the column of Isocrates.

23. Simms, "Trail for Archimedes's Tomb," 281–86, especially 283–84; Netz and Noel, *Archimedes Codex,* chap. 2.

24. Cf. Plut. *Marc.* 17.

25. For a *columella* as part of a deliberately understated tomb, see Cic. *Leg.* 2.66.

26. Recall how West and other artists portrayed the scene. See Simms, "Trail for Archimedes's Tomb," 281–86; Trapp, "Archimedes' Tomb," 286–88, with plates.

27. Steven White ("Cicero and the Therapists," in Powell, *Cicero the Philosopher,* 219–46) has drawn attention to the setting as that of Cicero's greatest and most recent grief. See also A. Erskine, "Cicero and the Expressions of Grief," in *The Passions in Roman Thought and Literature,* ed. S. M. Braund and C. Gill (Cambridge: Cambridge University Press, 1997), 36–47; M. Graver, *Cicero on the Emotions: "Tusculan Disputations" 3 and 4* (Chicago and London: University of Chicago Press, 2002), xii–xv. In "Home and Forum," Susan Treggiari discusses the stages of Cicero's grief in detail. On Cicero's withdrawal from active political life, see Habicht, *Cicero the Politician,* 68–86; Stockton, *Cicero,* 275–79.

28. These works are *Academica, De finibus, Tusculan Disputations, De natura deorum* (45 BCE); *De fato, Cato Maior de Senectute, Laelius de Amicitia,* and *De officiis* (44 BCE). On the relative weight of political and autobiographical reasons for Cicero's writing philosophy, see Schmidt, "Cicero's Place in Roman Philosophy," 121–23.

29. Cf. the beginning of the *De finibus,* which is set in 79 BCE. Cicero regarded his quaestorship in Sicily as a milestone in his development (*Brut.* 318.6): *cum autem anno post ex Sicilia me recepissem, iam videbatur illud in me, quicquid esset, esse perfectum et habere maturitatem quandam suam* (When, moreover, in the following year I had returned from Sicily, that quality in me, whatever it was, seemed then to have been completed and to have attained its full development).

30. On *exedi,* see Dougan and Henry, *Ciceronis Tusculanae Disputationes,* at 5.66. Cf. the use of *edax* in Hor. *Carm.* 3.30.3 and Ov. *Met.* 15.234, 872.

31. Once again this textual monument is particularly appropriate to its context. At the beginning of the *Tusculan Disputations* (1.5), Cicero says that the Greeks held geometry in the position of highest honor but that Romans have restricted the art to the purposes of measuring and reckoning: *in summo apud illos honore geometria fuit, itaque nihil mathematicis illustrius: nos metiendi ratiocinandique utilitate huius artis terminavimus modum.* Heine (Heine and Pohlenze, *Ciceronis Tusculorum Disputationum,* on 1.5) observes, "C. denkt etwa an Archimedes" ("Cicero is perhaps thinking of Archimedes"). The tomb marker, then, serves as a terminus marking the boundary between the two attitudes. On the Roman attitude toward mathematics, see Rawson, *Intellectual Life,* 156–69.

32. Authier ("Archimedes," 158) comments: "Beyond death, Cicero's account once more allows us to see the fundamental framework of Archimedes' thought, that of the effect of the minimum on the maximum: near the tomb we witness a socio-historical adaption of an experiment of Archimedes: all by himself, a simple inhabitant of Arpinum is responsible for the rediscovery of the monument to the extraordinary genius. This is undoubtedly the true resurrection of the man who, on the evening of defeat, fell in the dust with a compass in his hand."

33. See J. Graff, *Ciceros Selbstauffassung* (Heidelberg: Carl Winter, 1963), 46–54. Habinek ("Ideology for an Empire," 59) points to how the *re*-compounds in *Tusc.* 1.1 reinforce the idea of the return to philosophy.

34. On Cicero's approach to philosophical questions via history and traditional au-

thority, see Rawson, "Cicero the Historian and Cicero the Antiquarian," *JRS* 62 (1972): 33–45, especially 34–38.

35. *Tusc.* 1.61–65: *quae autem divina? vigere, sapere, invenire, meminisse.* See also 1.66–67, 70. On 1.62–64, see Frances Yates, *The Art of Memory* (Chicago: University of Chicago Press, 1966), 44–45. On ancient theories of memory in general, see Jocelyn P. Small, *Wax Tablets of the Mind: Cognitive Studies of Memory and Literacy in Classical Antiquity* (London and New York: Routlege, 1997); Vasaly, *Representations,* 100–102. On the memory evoked by monuments, see H. H. Davis, "Epitaphs and the Memory," *CJ* 53 (1958): 169–76.

36. Cicero also uses the memory of a representation to initiate philosophical discussions. At *Leg.* 1.1, Atticus recognizes the oak that he has read about often in Cicero's epic *Marius: lucus quidem ille et haec Arpinatium quercus agnoscitur saepe a me lectus in Mario.* At *Rep.* 6.10, the beginning of the *Somnium Scipionis,* Scipio recognizes the ghost of Africanus by its resemblance to his *imago*—his funerary bust: *Africanus se ostendit ea forma, quae mihi ex imagine eius quam ex ipso erat notior.*

37. In "Cicero's Use of Historical Examples," Brinton discusses Cicero's pairing the man of pleasure, Thorius, with the man of principle, Regulus: "The life of Regulus, we are to perceive, has a certain attraction which the life of Thorius lacks. Even though we might choose the existence of the former, we cannot help but admit that we would prefer the obituary of the latter" (177–78). As Virgil represents it (*Aen.* 6.660–65), having done what deserved a noble obituary gave entry to the Elysian Fields; see Habinek, "Science and Tradition," 231–38. See also Keith Hopkins, *Death and Renewal* (Cambridge: Cambridge University Press, 1983), 226–55.

38. Famous examples include the Scipio epitaphs, *Aen.* 6.660–65, and the *elogia* of the Augustan forum. See Roller, "Exemplarity," with bibliography.

39. On Dionysius's musical and literary proclivities, see Cic. *Tusc.* 5.63.

40. Or perhaps it poses a paradox. Michéle Lowrie draws my attention to *Arch.* 26: *ipsi illi philosophi etiam illis libellis, quos de contemnenda gloria scribunt, nomen suum inscribunt; in eo ipso, in quo praedicationem nobilitatemque despiciunt, praedicare de se ac nominari volunt* (For the philosophers themselves inscribe their own names on their very books about despising glory; and in the very writing in which they scorn praise and renown, they wish to be praised and cited by name).

41. As Cicero argues in *Rep.* 1.28, mathematical discovery is as memorable an achievement as holding public office.

42. On Caesar as tyrant, see Wood, *Cicero's Social and Political Thought,* 156–57.

43. Cic. *Att.* 12.15, 13.28. See D. L. Simms, "A Problem for Archimedes," *Technology and Culture* 30 (1989): 177–78. On Archimedes' defense of Syracuse, see Polyb. 8.6; Livy 24.34.1–16; Plut. *Marc.* 19. I discuss these passages in chap. 4 of the present study.

44. Douglas, "Form and Content in the *Tusculan Disputations*," 210–12.

45. Cicero commemorates Archimedes not only here but also when he talks about the Greek esteem for mathematics in *Tusc.* 1.5 and *Rep.* 1.28.

46. Habinek, "Science and Tradition," 237.

47. Edward Champlin, *Final Judgments: Duty and Emotion in Roman Wills, 200 B.C.–A.D. 250* (Berkeley and Los Angeles: University of California Press, 1991), 9 n. 16. Habinek ("Ideology for an Empire," 59) points to Cicero's argument that because the Greeks were unable to preserve their cultural capital, Rome had to take over the guardianship of it.

48. See Cic. *Leg.* 2.48, on the laws concerning the person responsible for performing the rites for the dead: *nulla est enim persona, quae ad vicem eius, qui e vita emigraret, propius accedat* (for there is no individual who comes closer to the position of the one who has departed life). On the heir's responsibility and the concern for the upkeep of the tomb, see Champlin, *Final Judgments*, 169–82.

49. On *inlustrare*, see Habinek, *Politics of Latin Literature*, 64–65. Cf. Livy pr. 10. My thanks to Habinek and to an anonymous reader for pointing out that by using the verb *excito*, Cicero is rousing the dead as well as calling Archimedes from his studies. See also Hor. *Carm.* 1.28.1–3, with R. G. M. Nisbet and M. Hubbard's extensive comments ad loc. in *A Commentary on Horace, "Odes," Book I* (Oxford: Clarendon, 1970).

50. See, e.g., *Tusc.* 1.13, where Cicero invokes the tombs of the Metelli and the Scipios to argue that the dead cannot not be wretched.

51. On a similar but more extended use of topography in philosophical dialogue, see C. Osborne, *Eros Unveiled: Plato and the God of Love* (Oxford: Clarendon, 1994), 86–116, especially 86–100.

52. Canter (*"Digressio* in the Orations of Cicero," 351 n. 1) points out that one of the uses Cicero identified for digression was to weaken or bury out-of-sight proofs on which the prosecution relies (Cic. *Part. or.* 5.15). A. Michel ("Rhétorique et Philosophique dans les 'Tusculanes,'" *REL* 39 [1961]: 164–6 n. 5) observes the blend of philosophical and traditional Roman authority. What John Dugan (*Making a New Man: Ciceronian Self-Fashioning in the Rhetorical Works* [Oxford: Oxford University Press, 2005], 104, says of the *De oratore* is appropriate here: "Ultimately the authority who guarantees the account that Cicero presents is Cicero himself: by undermining the autonomous reliability of *memoria* Cicero props up his own authority."

53. T. Murphy, "Cicero's First Readers: Epistolary Evidence for the Dissemination of His Works," *CQ* n.s. 48 no. 2 (1998): 505.

54. For superb discussion of this idea, see Dugan, *Making a New Man*, 6–13, 94–96.

55. See Cic. *Nat. D.* 1.6. At *Tusc.* 1.1, Cicero says that although he is returning to philosophy, the topic has been stored in his mind (*retenta animo*) all along.

56. When Cicero lists digression (*digressio*) among rhetorical figures in *De orat.* 3.203, he points out that its job is to please and then to return smoothly to the main topic at hand: *et ab re digressio, in qua cum fuerit delectatio, tum reditus ad rem aptus et concinnus esse debebit.*

57. Cicero's letters to Atticus about Tullia's shrine show that, earlier in the year, he had been very interested in memorials and the problems associated with them (see *Att.* 12.18, 19, 36). On tombs as reminders of the dead and of mortality, see Cic. *Tusc.* 1.31; *Sen.* 21 (Cato speaking, on old men's memories): *Nec sepulcra legens vereor, quod aiunt, ne memoriam perdam; his enim ipsis legendis in memoriam redeo mortuorum.* See also Varro *De lingua Latina* 6.49.

58. On the return to philosophy as a refuge rather than a pleasure (*delectatio*), see, e.g., Cic. *Fam.* 6.12.5, 9.2.5; for additional passages, see Graff, *Ciceros Selbstauffassung*, 133 n. 23. Cicero also enjoys one of the fruits of old age, remembering past achievements: cf. *Sen.* 71 (Cato speaking).

59. The anecdote fits into a particularly Roman view of argumentation, one in which "both a technical argument and an exhortation are appropriate for consideration of moral matters" (Habinek, "Science and Tradition," 247–48). Habinek need not have excepted the *Tusculans* from the philosophical works in which he observes this phe-

nomenon. Although employing dialogue, a mode of discourse that is untraditional from a Roman point of view, Cicero emphasizes memory and example in a way that is very Roman. See also M. Beard, "Cicero and Divination: The Formation of a Latin Discourse," *JRS* 76 (1986): 33–46.

60. See the extended development of this idea in the contemporary *De finibus* (5.1–8).

CHAPTER 3

1. On Cicero's rhetorical use of Marcellus, see chaps. 4–5 in the present study. On Cicero's use of examples from Roman history, see Dugan, *Making a New Man*, 11–12.

2. Cic. *Rep.* 1.21–22. On spheres in general, see A. Schlachter and F. Gisinger, "Der Globus: Seine Entstehung und Verwendung in der Antike" *Stoicheia* 8, Leipzig and Berlin, 1927; on mechanical planetaria like the second sphere, see 48–54. On these spheres, see Paul Keyser, "Orreries, the Date of [Plato] *Letter* ii, and Eudoros of Alexandria," *Archiv für Geschichte der Philosophie* 80 (1998): 241–67. Keyser argues that not even Archimedes had the kinematic astronomical theory or the technological method that would allow him to make a complicated device like the one attributed to him in *Rep.* 1.21. Such a device, Keyser argues (246), would have required differential gearing, not attested until the 80s BCE. Keyser (246–48) thinks that Cicero's source for the orrery is Posidonius, who was not expert in technology or astronomy and was known to exaggerate and present data tendentiously; Cicero welcomed Posidonius's attribution, because he needed such an orrery for G. Sulpicius Galbus to consult in 168 BCE. A complex bronze geared device found in a shipwreck off Antikythera in 1900 is dated to the second half of the second century BCE. The story of the original find is told in Derek Price, *Gears from the Greeks,* Transactions of the American Philosophical Society, n.s., vol. 64, pt. 7 (Philadelphia: American Philosophical Society, 1974), 5–10. For recent work on it and for images, see three pieces in *Nature* 444 (November 2006): J. Marchant, "In Search of Lost Time," 543–38; Fr. Charette, "High Tech from Ancient Greece," 551–52; T. Freeth et al., "Decoding the Ancient Greek Astronomical Calculator Known as the Antikythera Mechanism," 587–91, with bibliography.

3. On the multiple voices and the complexity of this dialogue, see M. Fox, "Dialogue and Irony in Cicero," in *Intratextuality: Greek and Roman Textual Relations,* ed. A. Sharrock and H. Morales (Oxford: Oxford University Press, 2000), 263–86. On the *De republica* in general, see M. Pohlenz, "Cicero *De re publica* als Kunstwerk," in *Festschrift Richard Reitzentstein,* ed. Eduard Fraenkel and Hermann F. Fränkel (Leipzig and Berlin: Teubner, 1931; reprinted in M. Pohlenz, *Kleine Schriften,* Hildesheim: G. Olms, 1965), 374–409; Zetzel, *Cicero, "De Re Publica,"* 1–29; K. Büchner, *De Re Publica: Kommentar* (Heidelberg: Carl Winter, 1984), especially 102–5 (on 1.21–22); M. Ruch, "La composition du *de republica,*" *REL* 26 (1948):157–71; P. L. Schmidt, "Cicero *De Re Publica:* Die Forschung der letzten fünf Dezennien," *ANRW* I.4 (1974): 262–333.

4. On Cicero's self-fashioning, see the important discussion by Dugan in *Making a New Man,* especially 21–74 (on the *Pro Archia*) and 75–171 (on the *De oratore*).

5. Writing to his brother, Cicero says that his friend Sallustius, after reading a draft of the first two books, urged him to give the dialogue a contemporary setting and speak in his own voice and with his own *auctoritas* (*Q. fr.* 3.5.1–2). Cicero chose to retain the setting but added prefaces addressed to his brother in his own voice. On this choice as a

way of reinforcing his *auctoritas*, see Murphy, "Cicero's First Readers"; Zetzel, *Cicero,* *"De Re Publica,"* 3. Dugan (*Making a New Man,* 88) and Dyck ("Cicero the Dramaturge," in *Qui Miscuit Utile Dulci: Festschrift Essays for P. Mackendrick,* ed. G. Schmeling and Jon O. Mikalson [Wauconda, IL: Bolchazy-Carducci, 1998], 154) see Sallustius's letter as an attempt to flatter Cicero, whose *auctoritas* had been compromised by the political situation. On using the voices of historical figures as a way of achieving an "ironic instability," see Fox, "Dialogue and Irony," 276–82.

6. On the fragmentary nature of the text, see K. Ziegler, ed., *M. Tullius Cicero: De Re Publica* (Leipzig: Teubner, 1958), iii–xiiii, with reference to the earlier editions; Büchner, *De Re Publica,* 16–20; Zetzel, *Cicero, "De Re Publica,"* 33; J. G. F. Powell, ed., *M. Tulli Ciceronis De Re Publica, De Legibus, Cato Maior De Senectute, Laelius De Amicitia* (Oxford: Oxford University Press, 2006), v–xxxii. On the implications of this fragmentation for interpretation, see Fox, "Dialogue and Irony," 270.

7. E. Gee, "Cicero's Astronomy," *CQ* n.s. 51, no. 2 (2001): 520–36; the phrase is from 534.

8. For interpretation of the phenomenon and discussion of Cicero's motives for using it, see M. Ruch, "Météorologie, Astronomie, et Astrologie chez Cicéron," *REL* 32 (1954): 200–219.

9. That Marcellus brought both spheres to Rome after the sack of Syracuse is a reasonable inference drawn from Cicero's words at *Verr.* 2.4.121: "We see what Marcellus brought to Rome at the Temple of Honor and Valour and likewise in other places." Polybius (9.10.3–13), the earliest source for the booty, does not mention spheres. Livy (25.40.1–3) refers in general to *ornamenta* from Syracuse and in particular to statues (*signa*) and paintings (*tabellae*). Livy says that foreigners used to visit the temples at the Porta Capena because of decorations (*ornamenta*) of this type (*eius generis*), but he does not mention spheres. In one version of Archimedes' death, Plutarch (*Marc.* 19.6) says that Archimedes was carrying spheres, among other instruments, to Marcellus. On the spoils in general, see M. Pape, *Griechische Kunstwerke aus Kriegsbeute und ihre öffentliche Ausstellung in Rome* (Diss. Hamburg: 1975), 6–7; J. J. Pollitt, "The Impact of Greek Art on Rome," *TAPA* 108 (1978): 155–74; A. W. Lintott, "Imperial Expansion and Moral Decline in the Roman Republic," *Historia* 21 (1972): 12–29; R. MacMullen, "Hellenizing the Romans (2nd Century B.C.)," *Historia* 40 (1991): 424–38; Gruen, *Culture and National Identity,* 84–130; Jaeger, *Livy's Written Rome,* 124–31; M. McDonnell, "Roman Aesthetics and the Spoils of Syracuse," in *Representing War in Ancient Rome,* ed. S. Dillon and K. E. Welch (Cambridge: Cambridge University Press, 2006), 79–105; M. McDonnell, *Roman Manliness: Virtus and the Roman Republic* (Cambridge: Cambridge University Press, 2006), 206–40.

10. The text printed is that of Powell, *M. Tulli Ciceronis De Re Publica.* Other discussions of this passage include those of A. Haury ("Cicéron et l'astronomie: À propos de *Rep.* 1.22," *REL* 42 [1964]: 198–212) and D. A. Kidd ("Notes on Aratus' *Phaenomena,*" *CQ* n.s., 31 no. 2 [1981]: 355–62), both of which are concerned with the astronomical reading of the passage. See also Emma Gee, "*Parva Figura Poli:* Ovid's *Vestalia* (*Fasti* 6.249–468) and the *Phaenomena* of Aratus," *PCPS* n.s., 43 (1997): 21–39; *Ovid, Aratus, and Augustus: Astronomy in Ovid's "Fasti,"* (Cambridge: Cambridge University Press, 2000), 96–100. Gee uses this passage and Cicero's description of a sphere in *Nat. D.* 2.88 to argue that Ovid takes on "Stoic baggage" by referring to Archimedes' sphere in *Fasti* 6.277–80.

11. My translation of the section describing the planetary sphere uses some expres-

sions from Robert Gallagher's more graceful rendering in "Metaphor in Cicero's *De Re Publica*," 510.

12. I follow Büchner (*De Re Publica*, 102), who thinks it more likely that the larger gap is after 1.22, not 1.23. For charts of possible arrangements of the folios, see Ziegler, *M. Tullius Cicero*, ix. See also Powell, *M. Tulli Ciceronis De Re Publica*, ad loc.

13. Büchner, *De Re Publica*, 102.

14. Gee ("Cicero's Astronomy," 527–36), using Cicero's *Aratea* as "a starting point for thinking about the connection of poetry, oratory, and natural philosophy in Ciceronian thought" (527–32), shows how a Stoic reading of the *Aratea* by the *De natura deorum*, in particular its discussion of celestial spheres and the movements of the stars, brings to light a parallelism between human and divine art, which, in turn, anticipates the analogy in the *De oratore* between the role of nature in the world and the role of the orator in verbal argument. Gee claims: "The primacy of words becomes for Cicero a philosophical position, one that enables him to articulate his own cultural location. Translation of Aratus is a good metaphor for the transfer of knowledge from Greece to Rome" (535). Gee also points outs that Cicero's dialogues create a "narrative of cultural transfer" (535–36), an observation that I develop later in the present chapter. In "Metaphor in Cicero's *De Re Publica*," Gallagher argues that the description of the spheres is a case of a "developed metaphor" (the term is Elaine Fantham's) for the political courses of states: "By expropriating the images of planetary motion and of the orrery in his discourse on states, Scipio makes concrete the abstraction of his political theory" (514–15). R. Coleman ("The Dream of Cicero," *PCPS* [1964]: 1–14) sees a pervasive Pythagoreanism linking the astronomical discussion and Scipio's dream.

15. *[S]phaeram quam M. Marcelli avus . . . sustulisset* (2) *. . . cuius . . . sphaerae* (2); *erat enim illa venustior et nobilior* (1); *coepit rationem huius operis. . . . Gallus exponere* (2); *sphaerae illius alterius solidae . . . eandem illam . . . cuius . . . ornatum* (1); *hoc autem sphaerae genus* (2); *in illa sphaera solida* (1); *[h]anc sphaeram Gallus cum moveret . . .* (2).

16. G. B. Conte (*The Rhetoric of Imitation: Genre and Poetic Memory in Virgil and Other Latin Poets*, trans. Ch. Segal [Ithaca: Cornell University Press, 1986], 36) describes this phenomenon in poetry: "a known poetic form or formula is conjured up, not simply to revive it by finding it a place in a new context but also to allow it to become the weaker member of a pair ('old' versus 'new') joined by a relationship of opposition or differentiation or a relationship merely of variation." The spheres, then, are joined by a relationship of differentiation (in degree of charm and fame) and of opposition (old and new).

17. See Gee, "*Parva Figura Poli*," 35. Spheres can be brought out in order to illustrate and to convince: in *De orat.* 3.162, Crassus, discussing metaphor, refers to a sphere that Ennius is said to have brought (*attulerit*) on stage. In *Nat. D.* 2.88, the speaker asks what would happen if someone brought (*tulerit*) the sphere recently made by Posidonius to Scythia or Britain.

18. Strabo (12.3.11) similarly tells of a sphere that Lucullus brought from Sinope after the sack of that city.

19. Schlachter and Gisinger, "Der Globus," 64–67. On globes as symbols of universal domination in late republican iconography, especially on coins, see C. Nicolet, *Space, Geography, and Politics in the Early Roman Empire* (Ann Arbor: University of Michigan Press, 1991), 35–37. Gee (*Ovid, Aratus, and Augustus*, 6) calls the sphere "a Roman appropriation of Greek astronomy" parallel to Ovid's use of Aratus in the *Fasti*.

20. On illustrious makers, see A. Brown, "Homeric Talents and the Ethics of Exchange," *JHS* 118 (1998): 166.

21. E. Rawson, *Roman Culture and Society* (Oxford: Oxford University Press, 1991), 582–98. Had Marcellus wanted weaponry to symbolize the conquest of Syracuse, he could have kept a siege engine, for, says Livy (26.21.7), catapults and ballistae were carried in his *ovatio*.

22. Gee (*Ovid, Aratus, and Augustus,* 6) sees a different shift: "it [the sphere] undergoes a metamorphosis in symbolic value by the context in which it is placed, in which political forces subsume the scientific."

23. Brown ("Homeric Talents," 165) discusses the way in which value follows values: "In the Homeric poems all valued items derive their value from the uses to which they are put by the heroic aristocracy. Since their way of life is dedicated at all levels to the pursuit of personal timē, all the property with which Homeric narrative is concerned is valued according to its perceived capacity to express that timē."

24. Gee (*"Parva Figura Poli,"* 35) suggests that Eudoxus's sphere could have been a treatise, for σφαῖρα can signify a globe, the heavens, and a treatise or a poem. As Gee has pointed out ("Cicero's Astronomy," 527–36), Gallus's account of the first sphere's development, which moves through the stages of physical sphere and then painted sphere/written treatise, conflates the physical item and the literature that describes or inspires it. Philus's account does not seem to be doing the same to Archimedes' planetary sphere, for which there was also an older textual account: Archimedes' sole mechanical treatise, according to Pappus of Alexandria (F. Hultsch, *Papp: Alexandrini Collectionis Quae Supersunt.* 3 vols. Berlin: Weidmann, 1876–78, book viii, p. 1026), was a work on sphere making. See Heath, *Works of Archimedes,* xvi.

25. Gallus is subject of the active verbs *iussisse proferri, coepit . . . exponere, dicebat,* and *moveret.*

26. Thus the sphere represents materially what Habinek ("Ideology for an Empire," 60) says of Greek philosophy (while discussing *Tusc.* 1.1): "Greek philosophy is a prestige-object that must be polished and put on display if it is to have its full impact as a marker of the status of its possessors."

27. See especially H. Berger Jr., "Levels of Discourse in Plato's Dialogues," in *Literature and the Question of Philosophy,* ed. A. J. Cascardi (Baltimore: Johns Hopkins University Press, 1987), 77–100.

28. On *venustior et nobilior in volgus,* Büchner (*De Re Publica,* 103) writes: "Symbol dafür, wie sich in einem Lernprozeß die geistige Elite von er Masse abhebt."

29. Rep. 1.14. A good parallel to the sphere passage is the story of Atlantis in Plato's *Timaeus.* Critias says that he heard it as a ten year old from his ninety-year-old grandfather, who heard it from his father (it seems), who heard it from his friend Solon, who heard it from the priests in Egypt. Notice also the specificity of the circumstances: Critias says that he heard it during the day of the Apaturia, which is called the Registration of Youth. On the implications of such historical precision for the interpretation of the *De oratore,* see Dugan, *Making a New Man,* 93; Dugan observes that Cicero displays his authority through his mastery of authentic historical detail.

30. Zetzel (*Cicero, "De Re Publica,"* 6) notes, "The transmission of knowledge and experience is simultaneously extolled and displayed."

31. E.g., Habinek, "Science and Tradition," 24–26. There are nine participants in the dialogue. The older, in addition to Scipio, are C. Laelius, L. Furius Philus, M'. Manilius,

and Sp. Mummius. The younger are Q. Aelius Tubero, nephew of Scipio; P. Rutilius Rufus, a protégé of Scipio and military tribune under him at Numantia; and Q. Mucius Scaevola and C. Fannius, both sons-in-law of Laelius.

32. The spheres embody the continuity that T. Murphy ("Cicero's First Readers," 505) observes in other dialogues: "Taken as a unit, the *Cato Maior* and the *Laelius* construct a vertical chain of transmission that extends down from Fabius Maximus and the time of the Second Punic War, through Cato Maior, Laelius, to Mucius Scaevola, and finally to Cicero, whose prefatory address to Atticus implies yet further disseminations." In addition, the importance of memory in the *De oratore* (see the discussion in Dugan, *Making a New Man*, 99–104) underscores its importance to Cicero's other dialogues.

33. The locus classicus on the obligation generated by gift exchange is M. Mauss, *The Gift: Forms and Functions of Exchange in Archaic Societies*, trans. Ian Cunnison (Glencoe: Free Press, 1954). See M. I. Finley, *Economy and Society in Ancient Greece*, ed. B. D. Shaw and R. P. Saller (New York: Viking, 1981), 233–45. See also A. Appadurai, ed. *The Social Life of Things: Commodities in Cultural Perspective* (Cambridge: Cambridge University Press, 1986), 41–43. A more recent overview is offered in C. A. Gregory "Exchange and Reciprocity," in *The Companion Encyclopedia of Anthropology*, ed. Tim Ingold (London and New York: Routledge, 1994), 911–39.

34. On the way in which setting a dialogue in the past helps establish "an aristocratic pedigree," see Dugan, *Making a New Man*, 81.

35. Scipio, too, says that the talk of two suns is not new: he and Rutilius used to discuss it under the walls at Numantia (*Rep.* 1.17).

36. *Rep.* 1.17: *et cum simul P. Rutilius venisset, qui est nobis huius sermonis auctor.* Of course, it would be impossible for Cicero to know any of the talk that went on before Rufus's arrival, since he claims him as his source. This is either a lapse in logic—if we are meant to receive this text as if it were the transcription of a conversation that actually took place—or a playful way of drawing attention to the dialogue's superimposition of textual reality on objective reality.

37. On Cicero as a *novus homo* fashioning himself per se, see Dugan, *Making a New Man*, 1–20, 94–96. Dugan (102) compares the assembly of interlocutors in the *De oratore* to the parade of *imagines* at an aristocratic funeral.

38. Usefulness is, of course, in the eye of the viewer. On items in Homer, Brown ("Homeric Talents," 165) writes, "Value in the Homeric economy, therefore, is related to utility, but not the utility of subsistence." I am grateful to my colleague José Gonzales for this reference. On Gallus's calming the troops, see A. C. Bowen, "The Art of the Commander and the Emergence of Predicative Astronomy," in Tuplin and Rihll, *Science and Mathematics*, 76–111.

39. Habinek, "Science and Tradition," 27. The utility of the sphere determines Scipio's verdict on Archimedes—that a good inventor contributes more to the state than does a tyrant (1.28): *Quis enim putare vere potest, plus egisse Dionysium tum cum omnia moliendo eripuerit civibus suis libertatem quam eius civem Archimedem cum istam ipsam sphaeram, nihil cum agere videretur, de qua modo dicebatur effecerit?* (For who, truly, can think that Dionysius accomplished more, when by all contrivances he snatched freedom from his fellow citizens, than Archimedes did when, although apparently accomplishing nothing, he produced that very sphere of which there was just talk?).

40. It is thus the more beautiful to the knowing eye. We may see, as Gee does in another connection ("Cicero's Astronomy," 530), an analogy with beauty in words, as in *De*

orat. 3.178: *sed ut in plerisque rebus incredibiliter hoc natura est ipsa fabricata, sic in oratione, ut ea, quae maximam utilitatem in se continerent plurimum eadem haberent vel dignitatis vel saepe etiam venustatis.*

41. As Cicero's Brutus says (*Brut.* 71), *nihil est enim simul et inventum et perfectum* (nothing is discovered and perfected all at once).

42. Zetzel, *Cicero, "De Re Publica,"* 16. In "Dialogue and Irony," Fox shows how the use of concrete examples to make theoretical points continuously undermines this dichotomy.

43. Gee (*"Parva Figura Poli,"* 26) observes, "The point Cicero's speaker is making about the more complex sphere built by Archimedes is that it supersedes the inventions of Thales and Eudoxus."

44. Rawson (*Intellectual Life,* 292) observes, "the Romans thought of Pythagoreanism as in a sense theirs, because it had been practised on Italian soil: the Greeks often called it 'the Italian philosophy.'"

45. Plato *Timaeus* 40b. Cicero translated this passage in his *Timaeus* (37): *si verbis explicare conemur nullo posito ob oculos simulacro earum rerum, frustra suscipiatur labor* (Should we attempt to explain these things with words, without an image of them placed before our eyes, our labor would be undertaken in vain). Both Plato and Cicero refer to the fear struck into the hearts of the ignorant (in Cicero, *rationis expertibus*) by eclipses. On Cicero and this dialogue, see C. Levy, "Cicero and the *Timaeus,"* in *Plato's "Timaeus" as Cultural Icon,* ed. G. Reydams-Schils (Notre Dame: University of Notre Dame Press, 2003), 95–110.

46. Polybius compares a comprehensive view of history to a view of an entire, living animal (1.4) and compares writing history without experience to drawing animals from stuffed dummies (12.25h).

47. Andrew Feldherr ("Cicero and the Invention of 'Literary' History," in Eigler et al., *Formen römischer Geschichtsschreibung,* 205) points out this double vision on the past and, noting that "the dialogue also measures itself against distinctively Roman historical practices," compares the introduction of the two suns at *Rep.* 1.15 and Cato's reference to eclipses as subject matter for the Annalistic tradition (H. Beck and U. Walter, ed. *Die frühen römischen Historiker,* vol. 1. 3 F 4.1).

48. Polyb. 9.14–20. Negative examples are Aratus's failed attack on Cynaetha, the attack of Cleomenes III on Polybius's hometown of Megalopolis, and Nicias at Syracuse. See Walbank, *Historical Commentary,* 2:138–49. See also Polyb. 3.36.6, with Walbank's commentary ad loc.

49. J. E. G. Zetzel ("Looking Backward: Past and Present in the Late Roman Republic," *Pegasus* 37 (1994): 20–30) makes the intriguing suggestion that Cicero wrote the *De republica* partly as a response to Lucretius's *De rerum natura.* The second sphere, then, would be even more effective as the symbol of an ordered cosmos.

50. The passage of the *Timaeus* referred to in the description of the second sphere is on the creation of time. Even the dramatic setting of the *Timaeus,* which takes place the day after the events of the *Republic,* is emblematic of the movement of time.

51. Gee ("Cicero's Astronomy," 535) claims that "Cicero is the new or updated Plato." Fox ("Dialogue and Irony," 276, with bibliography) sees the parallel between Scipio and Plato as a means by which "we become aware of all those later philosophers whose work contributed to Cicero's method."

52. On Thales as the greatest of the Seven Sages, see Cic. *Leg.* 2.26; Plut. *Solon* 4. For

testimonia on the Seven Sages, see H. Diels and W. Kranz, *Die Fragmenta der Vorsokratiker,* 6th ed., 3 vols. (Zurich: Weidmann, 1951; reprint, 1990–92), 1:61–66. For a recent interpretation of this tradition, see Richard P. Martin, "The Seven Sages as Performers of Wisdom," in *Cultural Poetics in Archaic Greece,* ed. C. Dougherty and L. Kurke (Oxford: Oxford University Press, 1998), 108–28; see also discussion in chap. 5 of the present study.

53. On Cicero's unfulfilled desire to see Egypt, especially Alexandria, see Cic. *Att.* 2.5.1–5; Rawson, *Cicero,* 108–9. On Archimedes in Egypt, see Diodorus 5.37.3–4, who says that Archimedes invented the water screw there. On Thales, see Diogenes Laertes 1.22–24 (citing Pamphilus), 1.27 (Thales is said to have measured the pyramids); Diels and Kranz, *Fragmenta der Vorsokratiker,* 1:68. On Pythagoras, see Isocrates, *Busiris* 28, on which see Kahn, *Pythagoras and the Pythagoreans,* 5–12. Whether or not these figures really traveled to Egypt is of less importance for our purposes than the tradition that they did. On Sicily and the complex issue of what it was to be Sicilian in a Roman world, especially in the eastern half, see K. Lomas, "Between Greece and Italy: An External Perspective on Culture in Roman Sicily," in Smith and Serrati, *Sicily from Aeneas to Augustus,* 161–73.

54. Zetzel, *Cicero, "De Re Publica,"* 6. The same goes for what remains of Cicero's preface, which is also framing material.

55. Feldherr ("Cicero and the Invention of 'Literary' History," 205–6) points out how the dialogue is sparked by the *prodigium* and how the talk of it places emphasis on personal authority.

56. Cic. *Verr.* 2.4.115: *ab illo qui cepit conditas, ab hoc qui constitutas accepit captas dicetis Syracusas* (You will say that Syracuse was founded by him who captured it, captured by him who received it when it had been established). At Livy 31.31.8, when the Roman *legatus* counters Philip's charges during the Aetolian congress, he says that the Romans brought aid to the Syracusans, who were suffering grievously under foreign despots, but that since the Syracusans themselves prefered to be enslaved to despots rather than be captured by the Romans, "we restored to them their city, captured and freed by the same arms" (*iisdem armis et liberatam urbam reddidimus*).

57. Traces of such a tradition may remain in Plutarch's report (*Marc.* 19.6) that Archimedes was bringing the spheres to Marcellus when he fell in with Roman soldiers who killed him because they thought that he was carrying gold. On this passage, see chap. 4 in the present study. J. Geiger ("Contemporary Politics in Cicero's *De Republica,*" *CP* 79 [1984]: 38–43) suggests that the reference to Marcellus's restraint was meant as a parallel to the story that Cato took from Cyprus only a statue of Zeno, founder of Stoicism; Keyser ("Orreries," 248 n. 34) notes that this restraint in famous generals is a topos. See Val. Max. 4.3 on "Moderation." A famous example is Aemilius Paullus's taking only the library of Macedon; see Plut. *Aem.* 28.10; Cic. *Off.* 2.76.

58. See Dugan, *Making a New Man,* 99–103, on Cicero's connection of memory to trauma in the *De oratore.*

59. Kept as a prestige item and shown only to select individuals, the sphere is no more useful than the tripod that made the rounds of the Seven Sages and returned to Thales in the end. The circulation of the tripod did the Sages good, because it increased their renown (Plut. *Solon* 4). I owe this reference to David Elmer.

60. My thanks to Andrew Feldherr for drawing my attention to this.

61. Thus this passage anticipates Cicero's claim to have pointed out Archimedes'

tomb to the locals. Cf. *Tusc.* 5.65, where Cicero said that he immediately (*statim*) told the local dignitaries, who were with him, that he had found what he sought. He did not keep his discovery of Archimedes' tomb to himself but handed it on to a group of Syracusans.

62. Habinek, *Politics of Latin Literature,* 45–68. Habinek (67) argues, "By expanding the commonwealth of the Romans and making that wealth available to later and larger groups of readers through his extensive literary output, Cicero succeeded in widening the circle of self-conscious, interconnected elites . . . and in making it possible for them to believe, regardless of their status or place of origin, that they too had a stake in the preservation of a distinctively Roman *mos maiorum.*"

63. Habinek (*Politics of Latin Literature,* 102) writes, "If circulating a written text, as opposed to delivering an oral performance, increases an author's renown by extending it in both space and time, it also runs the risk of disconnecting an author from his text and undercutting the importance of personal presence." See also Dugan, *Making a New Man,* 43–47.

64. On this feature of Archimedes' death, see chap. 4.

65. *Rep.* 6.12: *Hic cum exclamasset Laelius ingemuissentque vehementius ceteri: "St! Quaeso," inquit, "Ne me ex somno excitetis et parumper audite cetera!"*

66. On the risk inherent in gift giving (that the gift will be in vain), see Finley, *Economy and Society,* 237–38.

67. *OLD,* s.v. *exponere.* See also Scipio's words at Cic. *Rep.* 1.46.70: *Simul et qualis sit et optimum esse ostendam, expositaque ad exemplum nostra re publica, accommodabo ad eam si potuero omnem illam orationem quae est mihi habenda de optimo civitatis statu.*

68. At Cic. *Rep.* 2.4, Scipio says that the story of Romulus's divine parentage was both old and "wisely passed down by our forefathers" (*sapienter a maioribus prodita*). On the regal period in *Rep.* 2, see M. Fox, *Roman Historical Myths* (Oxford: Oxford University Press, 1996), 5–28.

69. *OLD,* s.v. *exponere* 6.b; Cic. *De orat.* 1.227; Cic. *Tusc.* 1.21. On the nature of book releases and distribution in republican Rome, see Murphy, "Cicero's First Readers," 495–96, 499–501, with bibliography.

70. Feldherr ("Cicero and the Invention of 'Literary' History," 206) observes another similarity between the dialogue and the real Republic: "neither is allowed to be the invention of one man but the product of generations of statesmen remembering and recalling the *instituta* of their ancestors, as Cicero here remembers Rutilius Rufus, remembering Scipio, remembering Cato, remembering Romulus."

71. Cic. *Off.* 1.1–8.

72. Fox ("Dialogue and Irony," 272) observes: "In an important sense, argumentative strategies themselves become subjected to a distancing ironic viewpoint, and come to take on the same kind of role as the fictional personages who inhabit the dialogue. Arguments become symbols; they do not lose their explicit purpose, but they acquire a further dimension, and begin to represent ways of thinking which are seen to be available, suggestive, even productive, but which are not necessarily validated through explicit presentation by Cicero." These interruptions are less argumentative strategies than they are symbols of other disruptions.

73. Examples include Aristophanes' hiccups and Alcibiades' interruption in the *Symposium,* a dialogue presented as an account of a banquet related years after the event, with many layers of narrative embedding. See J. Henderson, "The Life and Soul of

the Party," in Sharrock and Morales, *Intratextuality,* 287–324. The *Timaeus,* too, has many layers of embedding. (See, in the present chap., n. 29, on the story of Atlantis.)

74. Archimedes' creation of the two spheres also anticipates what Dugan (*Making a New Man,* 199–201) calls "the complementary phenomena" of *aemulatio* and *imitatio* that drive Roman rhetorical history in the *Brutus.*

75. Feldherr, "Cicero and the Invention of 'Literary' History," 206.

CODA TO PART ONE

1. Firmicus wrote this treatise during the period 334–37 CE. The text is from the Teubner edition, vol. 1, W. Kroll and F. Skutsch, eds., *Firmicus Maternus, Mathesis,* (Leipzig: Teubner, 1897). W. Kroll, F. Skutsch, and K. Ziegler, eds., *Firmicus Maternus, Mathesis,* vol. 2 (Stuttgart: Teubner, 1968). See also P. Monat, ed. and trans., *Mathesis/Firmicus Maternus* (Paris: Les Belles Lettres, 1992). The last ancient astrological text to be written in Latin, the *Mathesis* has been called, by a translator, "one of the last great statements of the thoughts and feelings of pagan Rome" (Jean Rhys Bram, trans., *Ancient Astrology: Theory and Practice = Matheseos Libri VIII* [Park Ridge, NJ: Noyes, 1975], 2). It was written between 337 and 354 CE. See T. Barton *Ancient Astrology* (London and New York: Routledge, 1994), 80.

2. See chap. 4. Archimedes' horoscope appears at *Mathesis* 6.30.26, following those of Demosthenes, Homer, Plato, Pindar, and Archilochus.

3. Bram, *Matheseos Libri VIII,* 303.

4. Bram, *Matheseos Libri VIII,* 2.

5. The whole phrase is *honestas et varias sermonum fabulas serebamus,* which Bram (*Matheseos Libri VIII,* 11) translates as "exchanging literary references" and which Monat (*Mathesis/Firmicus Maternus,* 50) translates more closely as "un conversation portant sur des sujets aussi élevés que variés." Monat (11) observes that book 1 begins "un peu à la manière d'un dialogue cicéronien, par le rappel d'une conversation entre l'auteur et son protecteur."

6. Firmicus refers to Sicily as the land "that I inhabit and whence I come by descent" (*quam incolo et unde oriundo sum*). The words *civis meus* in Archimedes' horoscope (*Mathesis* 6.30.26), which suggest that Firmicus, too, was a Syracusan, are Skutsch's conjecture.

7. Cf. Cic. *Rep.* 1.22: *plus in illo Siculo ingenii quam videretur natura humana ferre potuisse iudicavi fuisse* (I decided that there was more genius in that Sicilian than human nature seemed able to admit).

8. Barton, *Ancient Astrology,* 136–37. The preface to book 7 presents astrology as if it were a mystery religion, refers to Orphic and Pythagorean initiation, and demands of Mavortius an oath that he will not disclose to profane ears what he has learned. On the passage, see Bram, *Matheseos Libri VIII,* 313 n. 79.

9. Cf. *Mathesis* 5 pr. 5 (trans. Bram, *Matheseos Libri VIII,* 156): *ut quicquid divini veteres ex Aegyptiis adytis protulerunt, ad Tarpeiae rupis templa perferret* (Our purpose is to convey to the temples of the Tarpeian Rock whatever the divine ancients of Egypt brought forth from their shrines).

10. The fiction that the work was requested by the dedicatee is common in classical times and becomes even more common in late antiquity (see Janson, *Latin Prose Prefaces,*

116–17). Expressions of authorial incompetence and unworthiness are also common in classical times and, according to Janson (124–25), "enormously common" in late Latin.

11. E.g., *Mathesis* 2.30.15: *adgredere hoc opus et posteriores hos libros memoriae trade, ut integram tibi scientiam divinitas tradat . . .* ; 8.33: *in ceteris vero libris Romanis novi operis tradidimus disciplinam.* Mavortius is to hand this knowledge on to his children and friends (*filiis tuis trade; tuis trade amicis,* 8.33.3).

12. *Mathesis* 3 pr. 1 (*Matheseos sermo totus . . . in sententias transferatur*), 5.7.5, 6.40.4, 7.26.12 ("But now that these things have been explained, let our entire discourse be shifted [*transferatur*] to the explanation [*expositionem*] of the Barbarian Sphere").

13. *Mathesis* 4 pr. 4, 5 pr. 1.

14. Josèphe-Henriette Abry, "Manilius et Julius Firmicus Maternus, Deux Astrologues sous L'empire," in *Imago Antiquitatis: Religiones et Iconographe du Monde Romain, Mélanges offerts à Robert Turcan,* ed. N. Blanc and A. Buisson (Paris: De Boccard, 1999), 35–45. See p. 35.

15. After all, as Firmicus says repeatedly in book 1, the Sicilians are clever, while the Romans are suited to ruling. When Firmicus and Maternus converse, Roman governing talent meets Sicilian intelligence. See also *Mathesis* 5 pr. 4 (trans. Bram, *Matheseos Libri VIII,* 156), an invocation to the supreme deity: *Pura mente et ab omni terrena conversatione seposita et cunctorum flagitiorum labe purgata hos Romanis tuis libros scripsimus, ne omni disciplinarum arte translata solum hoc opus extitisse videatur, ad quod Romanum non adfectasset ingenium* (With purified soul, freed from earthly contacts and all taint of sin, we have written this book for Thy Romans in order that this subject may not be the only one which the Roman genius has not pursued).

A SKETCH OF EVENTS AT SYRACUSE

1. Polyb. 7.2–8, 8.3–7; Cic. *Verr.* 2.4.115–31; Cic. *Off.* 5.50; Cic. *Rep.* 1.21–22; Livy 24.4–7, 21–33, 25.23–31, 25.40; Plut. *Marc.* 14–22. For a brief overview with references to the sources, see Briscoe, "Second Punic War," 61–62; for more detailed analysis, see Eckstein, *Senate and General,* 135–65, with a chronological appendix at 345–49; for the chronology, see also Walbank, *Historical Commentary,* 2:5–8.

CHAPTER 4

1. Valerius dedicated his work to the emperor Tiberius (14–37 CE). For discussions of Valerius with differing opinions as to the audience and purpose of his work, see Bloomer, *Valerius Maximus;* Clive Skidmore, *Practical Ethics for Roman Gentlemen: The Work of Valerius Maximus* (Exeter: University of Exeter Press, 1996); Alain Gowing, *Empire and Memory: The Representation of the Roman Republic in Imperial Culture* (Cambridge: Cambridge University Press, 2005), 49–62.

2. (*Factorum et dictorum memorabilium libri IX* 8.7.ext.7; text from J. Briscoe, ed., *Valerius Maximus: Factorum et Dictorum Memorabilium Libri IX,* vol. 2 [Stuttgart and Leipzig: Teubner, 1998]; see also D. R. Shackleton-Bailey, ed., *Valerius Maximus, "Factorum et Dictorum Memorabilium Libri IX,"* 2 vols. [Cambridge, MA: Harvard University Press, 2000], vol. 1, with general bibliography).

3. Clive Skidmore (*Practical Ethics,* xi) says that there are more surviving medieval and Renaissance manuscripts of Valerius Maximus than of any other ancient prose au-

thor. For a collection of medieval Archimedes narratives (from the mid-thirteenth through mid-fourteenth centuries), see Clagett, *Archimedes in the Middle Ages*, 1329–41. According to Clagett, the only event in Archimedes' life most medieval biographers knew of and thought significant was his death. They relied almost entirely on Valerius, with a few noting material from Orosius (on the death) and Lactantius (on the sphere). Representations of the death scene in art include a Renaissance mosaic once thought to be ancient (for images, see Franz Winter, *Der Tod des Archimedes* [Berlin: De Gruyter, 1924]) and several bookplate engravings (see, e.g., Mazzuchelli, *Notizie*, 1, where the soldier uses a backhanded stroke). For a list of images, see Andor Pigler, *Barockthemen: Eine Auswahl von Verzeichnissen zur Ikonographie des 17. und 18. Jahrhunderts*, 3 vols. (Budapest: Akadémiai Kiadó, 1974), 2:368–69.

4. Skill at making smooth and clever transitions, Bloomer points out (*Valerius Maximus*, 25), is a result of extensive training in declamation: the clever transition draws attention to itself and to "the virtuosity of the speaker who employs it."

5. There are seven exempla but eight individuals, since the senator Publilius and the knight Lupus Pontius, both of whom persevered in forensic activities even after losing their eyesight, share an exemplum (8.7.5). Valerius comments on several other figures who, like Archimdes, worked and learned right up to the moment of death: Terentius Varro among the Romans (8.7.3); among the foreigners, Plato, Carneades, Chrysippus, Cleanthus, Sophocles, and Solon. Plato was said to have had the mimes of Sophron under his head as he died (8.7.ext.3); Carneades worked until the moment of death (8.7.ext.5); Chrysippus wrote in his old age (8.7.ext.10); Cleanthus taught until he was ninety-nine (8.7.ext.11); Sophocles wrote *Oedipus at Colonus* while dying (8.7.ext.12); Solon, who grew old learning something every day, was learning the day he died (8.7.ext.14).

6. For another contrafactual introduction, see 6.1.6. Valerius's use of the agricultural metaphor *fructuosa* for Archimedes' *industria* may have been inspired by the preceding exemplum, that of Anaxagoras, who neglected cultivating his estates while traveling abroad to further his education and said, "I should not be alive had these not perished." On the order of the collection and the connecting sentences in general, see F. R. Bliss, "Valerius Maximus and His Sources: A Stylistic Approach to the Problem" (PhD diss., University of North Carolina at Chapel Hill, 1951), 9–18.

7. Compare Valerius's use of active verbs in the stories of killings that are a point of pride: Lucretia, for example, "killed herself with the sword" (*ferro se . . . interemit*, 6.1.1); Verginius "led his daughter into the forum and killed her" (*deductam in forum puellam occidit*, 6.1.2); Pontius Aufidianus "even killed his own daughter" (*etiam ipsam puellam necavit*, 6.1.3); and Horatius "killed . . . his own sister . . . with a sword" (*sororem suam. . . . gladio . . . interemit*, 6.3.6). Gowing (*Empire and Memory*, 49–62) discusses Valerius's selectivity, on both the large and small scale: Valerius draws his exempla from the period before Actium; his removal of a given story from its historical context allows him to tweak its meaning; he also on occasion pointedly omits a detail in order to erase a memory, as when, rebuking Sejanus, he fails to mention him by name. The instability and flexibility of exemplary discourse is emphasized in Roller, "Exemplarity," 23, 51.

8. The soldier with the sword, the slaughtered man, and the blood in the dirt would fall into Quintilian's second class of signs, those that are unnecessary (*eikota*); within this class, they would fall into the third type, those consistent with common sense (*nonrepugnans*) for example: *in doma furtum factum ab eo qui domi fuit*, "if there has been a

theft in the house it was done by a person in the house" (Quintitian, *Inst.* 5.10.16). On ancient classifications of signs, see G. Manetti, "The Concept of the Sign from Ancient to Modern Semiotics," in *Knowledge through Signs: Ancient Semiotic Theories and Practices,* ed. G. Manetti (Bologna: Brepols, 1996), 28–31. In *Theories of the Sign in Classical Antiquity,* trans. Christine Richardson (Bloomington and Indianapolis: Indiana University Press, 1993), 139–56, Manetti cites the author of the *Rhetorica ad Herrenium,* who writes of the *constitutio coniecturalis,* in which signs are sought to show whether or not the accused individual performed the criminal deed.

> Ajax, in the forest, after he recognized what he had done in his madness, fell upon his sword. Ulysses comes upon him, sees him dead, and draws the bloody weapon from the body. Teucer comes upon him, sees his brother dead and his brother's enemy with the bloody sword. He charges Ulysses with a capital crime. (*Rhet. Her.* 1.18)

Valerius's readers—especially if students of declamation, as Bloomer argues in *Valerius Maximus* (14–17)—would be conscious of making the same leap to a conclusion as did Teucer, although in Archimedes' case it was correct; they would likewise be sensitive to the possible implications of a writer's omitting such a detail as the actual killing blow. A good example of such omission in practice is Lysias 1.26–27, where the defendant is careful to leave out his actual killing of Eratosthenes.

9. Valerius personifies Industria strikingly in the opening passage of this section (8.7.1): he says that military service is fortified "by her keen spirit" (*alacri spiritu*) and that "all studies are received into and nourished in her faithful bosom" (*fide sinu cuncta studia recepta nutriuntur*).

10. Even Valerius's preface presents this entire project as that of a collector, who attributes the authority for his stories to the sources from which he draws them: "I have decided to select from famous authors and set out in an organized way [*ab inlustribus electa auctoribus digerere constitui*] deeds and sayings alike worthy of remembrance, of both the city Rome and foreign nations—[deeds and sayings] so broadly spread among other authors that they cannot be learned briefly—so that those wanting to take up examples might avoid the effort of a lengthy inquiry." At 1.8.7, he points out that they are "handed down" (*tradita*), not "new" (*nova*). See Gowing, *Empire and Memory,* 55. In the account of Archimedes' death reported by Zonaras and Tzetzes and attributed to Cassius Dio, the narrating Dio may also have taken responsibility for the story. Yet this version, too, suggests that Archimedes provoked the soldier and thus brought about his own death (see discussion later in the present chap.).

11. Called *De signis* (On the Statues) because it catalogs the artwork stolen by Verres when praetor in Sicily, this speech was never delivered in court. Verres went into exile after the preliminary hearing, so Cicero published *Verr.* 2.4 together with the other speeches of the second *actio* as a rhetorical tour de force. On this speech and its use of artwork, see Vasaly, *Representations,* 104–30. For a perceptive analysis of the speech's effect on later discussions of plunder and appropriation, see M. Miles, "Cicero's Presentation of Gaius Verres: A Roman View of the Ethics of Acquisition of Art," *International Journal of Cultural Property* 2 (2002): 28–49. On the role of this speech in Cicero's self-fashioning, see Dugan, *Making a New Man,* 8–12.

12. Roughly the last quarter of the speech (2.4.115–49) is devoted to Verres' behavior in Syracuse. Syracuse concludes the section devoted to thefts from several towns

(2.4.75–135). Vasaly (*Representations*, 122–23 and n. 50) points out how the topos of the "captured city" overlaps with the portrait of Verres as the stereotypical, lust-ridden tyrant.

13. *Scientiam pollicentur, quam non erat mirum sapientiae cupido patriae cariorem <esse>. Atque omnia quidem scire, cuiuscumque modi sint, cupere curiosorum, duci vero maiorum rerum contemplatione ad cupiditatem scientiae summorum virorum est putandum* (text from L. D. Reynolds, ed., *Cicero, "De Finibus"* [Oxford: Oxford University Press, 1998]).

14. Pliny *N.H. 7.125: Grande et Archimedi geometricae ac machinalis scientiae testimonium M. Marcelli contigit interdicto, cum Syracusae caperentur, ne violarentur unus, nisi fefellisset imperium militaris inprudentia.*

15. *Mathesis* 6.30.26: *. . . qui Romanos exercitus mechanicis artibus saepe prostravit. Hunc Marcellus in triumpho victoriae constitutus, inter ovantes militum strepitus et triumphales laureas collocatus, lugubri maerore deflevit.*

16. Plut. *Marc.* 19.4–6. The translations of the three passages in the text are modified from B. Perrin, trans., *Plutarch's Lives*, vol. 5 (London: W. Heinemann; New York: G. P. Putnam's Sons, 1917).

17. Polyb. 8.3.1–7.9; Livy 24.34.1–16; Plut. *Marc.* 14.1–18.2.

18. The surviving text of Polybius's account breaks off with the Romans' initial entrance into the city (8.37.5–13).

19. Zonaras 9.5: πὰρ κεφαλάν, ἔφη, καὶ μὴ παρὰ γραμμάν.

20. Tzetzes *Chil.* 2.142: τὶ μηχάνημά τις τῶν ἐμῶν μοι δότω.

21. Roman criminal courts would not have categorized charges this way. Andrew Riggsby (*Crime and Community in Ciceronian Rome* [Austin: University of Texas Press, 1999], 200 n. 2) points out that "Roman law does not have a notion comparable to 'malice aforethought' which would distinguish murder from other forms of homicide." The rationalizations of Zonaras and Tzetzes resemble more the various positions taken in rhetorical argument. See, e.g., the accusations and rebuttals in Quintilian *Inst.* 3.17–19.

22. Roller ("Exemplarity," 9) describes the attempt to recover the historical actualities behind exempla and observes that it is thus argued "that the stories of Horatius Cocles, Cloelia, and Mucius Scaevola provided a fig leaf by which the Romans dissembled the fact that Porsenna captured the city." Archimedes' death could also be seen as one of those events that as Benedict Anderson observes, people are obliged "to remember/forget," such as the fact that the American Civil War was between two sovereign entities. See Anderson, *Imagined Communities: Reflections on the Origin and Spread of Nationalism*, 2nd ed. (London and New York: Verso, 1991), 199–203, esp. p. 201.

23. The killer is the antithesis of the shipwrecked man of *De republica* 1.29, who, seeing geometrical diagrams drawn in the sand on the shore, rejoiced that he was near civilization, for he saw the traces of men (*hominum vestigia*). Roman defensiveness about mathematical inferiority vis-à-vis the Greeks is represented by Cic. *Tusc.* 1.5. See also Rawson, *Intellectual Life*, 156–69.

24. Tzetzes *Chil.* 2.148: τὸν δὲ φονέα τοῦ ἀνδρὸς οἶμαι πελέκει κτείνει. Plutarch and Tzetzes—the one saying that Marcellus treated the killer as if he were polluted, the other that Marcellus had him beheaded—put their fingers on it nicely: the soldier is a scapegoat.

25. The killer's anonymity brings to mind Aeneas's thoughts about killing Helen at *Aen.* 2.583–84: *namque etsi nullum memorabile nomen feminea in poena est, nec habet*

victoria laudem . . . (For even if there is neither memorable fame to be won in punishing a woman, nor praise for such a victory . . .). On this doubtful passage, see C. Murgia, "The Date of the Helen Episode," *HSCP* 101 (2003): 405–26, with full bibliography.

26. *Verr.* 2.4.115. Of course, we have to take Cicero's word for the story's familiarity. But if it were totally unfamiliar, he would lose credibility, and his contrast would hardly be effective.

27. This question is asked by Habinek ("Ideology for an Empire," 60) in a similar context.

28. Laird, *Powers of Expression*, 16–17.

29. Scenes of individuals and groups receiving news allow writers to illustrate the character of the recipient, the nature of the event, or both. The messenger speech, of course, is central to tragedy.

Consider how the historian Livy, for example, uses this technique to illustrate character. On the one hand, he describes the consul Horatius continuing to carry out the dedication of the temple of Jupiter on the Capitoline, even when told that his son had died: "whether he did not believe that it had happened or was of such steadfastness of spirit [*tantum animo roboris fuerit*], we are not told for certain, nor is it easy to determine" (2.8.8); on the other hand, the immediate Roman response to the news of the disaster at Cannae illustrates the devastating nature of the event: "Never, with the city safe, was there such fear and uproar within the walls of Rome. And so, I shall give way under my burden, and not attempt to describe what I shall diminish by telling" (22.54.7–8). See also Val. Max. 5.10.1. On the reaction to Cannae, see Jaeger, *Livy's Written Rome*, 99–107.

30. On the life of Marcellus, see F. Münzer, "Claudius," *RE* 3.2:2738–55; E. Carawan, "The Tragic History of Marcellus and Livy's Characterization," *CJ* 80 (1984–85): 131–41. On Clastidium, see Livy *Periochae* 20; Plut. *Marc.* 6.6–7.8; Val. Max. 3.2.3; Prop. 4.10.33–44.

31. On Marcellus's contested reputation, see Harriet Flower, "'Memories' of Marcellus: History and Memory in Roman Republican Culture," in Eigler et al., *Formen römischer Geschichtsschreibung*, 39–52. Flower places the criticism of Marcellus in the context of the various systems of memory taking shape in Rome during the late Republic and observes: "Denigration formed a kind of dialogue with the positive traditions that celebrated Marcellus as a great Roman hero, worthy of remembrance and of imitation" (40).

32. Who were his enemies? Livy (26.26.6–9) has Marcellus claiming that M. Cornelius Cethegus encouraged the Sicilians to complain about his treatment of them; Flower ("'Memories,'" 47) points out that the success of Syracuse, when juxtaposed to the defeat of the Scipios in Spain, would have increased the hostility between his gens and the patrician Cornelii. Livy (34.4.3) has Cato attacking the artwork. For the attacks on Marcellus, see Eckstein, *Senate and General*, 169–77; for conflicts between Marcellus and the Fabii, see McDonnell, *Roman Manliness*, 206–40.

33. Polybius omits Marcellus's victory in single combat at Clastidium and the *spolia opima*. He criticizes both Marcellus's taking the Syracusan art and the improvidence that led to his untimely death (10.32–33). See Flower, "Memories," 46–51; McDonnell, *Roman Manliness*, 208.

34. Valerius is quite explicit in 8.7.ext.7: Marcellus felt that almost as much glory lay in saving Archimedes as in crushing Syracuse.

35. *Mathesis* 6.30.26: *in triumpho victoriae constitutus, inter ovantes militum strepitus et triumphales laureas collocatus.* Marcellus observed two ovations in the summer of 211, one in Rome and the other on the Alban mount (which Livy calls a triumph).

36. Livy (26.21.7) refers to "catapults, machines for discharging projectiles, and all the other implements of war" (*catapultae ballistaeque et alia omnia instrumenta belli*).

37. On Marcellus as patron, see Eckstein, *Senate and General*, 171–77, 183–84; on later Marcelli, see Hor. *Carm.* 1.12.46, with Nisbet and Hubbard, *Commentary on Horace*, ad loc. Flower ("Memories," 41) observes that Marcellus's role as ancestor of the Marcellus who became nephew and son-in-law of Augustus renewed his status at the start of the principate.

38. See G. Maslakov, "Valerius Maximus and Roman Historiography," *ANRW* II.32.4 (1986): 443–45, 453–57, on the decontextualizing of exempla; see, more recently, Roller, "Exemplarity," 31–32.

39. See John Marincola, "Marcellus at Syracuse (Livy xxv.24.11–15): A Historian Reflects," in *Studies in Latin Literature and Roman History*, ed. C. Deroux, Collection Latomus 287 (Brussels: Latomus, 2005), 12:219–29; A. Rossi, "The Tears of Marcellus: History of a Literary Motif in Livy," *Greece and Rome* 47, no. 1 (2000): 56–66; Jaeger, "Livy and the Fall of Syracuse," in Eigler et al., *Formen römischer Geschichtsschreibung*, 230–34.

40. Cic. *Verr.* 2.4.115: *Vnius etiam urbis omnium pulcherrimae atque ornatissimae, Syracusarum, direptionem commemorabo et in medium proferam, iudices, ut aliquando totam huius generis orationem concludam atque definiam.*

41. *Nemo fere vestrum est quin quem ad modum captae sint a M. Marcello Syracusae saepe audierit, non numquam etiam in annalibus legerit.* B. W. Frier (*Libri Annales Pontificum Maximorum: The Origins of the Annalistic Tradition*, Papers and Monographs of the American Academy in Rome 27 [Rome: American Academy in Rome, 1979; reprinted with a new introduction, Ann Arbor: University of Michigan Press, 1999], 221) interprets these annals to be not a specific author's work but "a single tradition, in effect, the public history of Rome."

42. For the horrific images that the word *direptio* would have brought before the mind's eye of the audience, see A. Ziolkowski, "*Urbs Direpta*, or How the Romans Sacked Cities," in *War and Society in the Roman World*, ed. J. W. Rich and G. Shipley (London and New York: Routledge, 1993), 69–91.

43. Livy 25.31.9. See Ziolkowski, "*Urbs Direpta*," 82–83, on the "cold-blooded hypocrisy" with which Marcellus violated the conditional surrender of Achradina and Nasos. On the art, see McDonnell, *Roman Manliness*, 228–35, with bibliography.

44. See Cic. *Verr.* 2.4.86–87 (on C. Marcellus), 89–91 (on M. Marcellus Aserninus). On the audience, see P. A. Brunt, "Patronage and Politics in the *Verrines*," *Chiron* 10 (1980): 273–89. Cicero seems to have embraced the challenge eagerly. Although this speech was never delivered in court, it demonstrated his skill at handling such material.

45. See C. E. W. Steel, *Cicero, Rhetoric, and Empire* (Oxford: Oxford University Press, 2001), 46–47, on Cicero's caution when it came to Greek material. Steel notes that even the discovery of Archimedes' tomb could have been fodder for attack from a competent orator. Marcellus, then, provides Cicero a precedent for philhellenism, but one set safely in the past.

46. Cic. *Verr.* 2.4.115: *Conferte hanc pacem cum illo bello, huius praetoris adventum cum illius imperatoris victoria, huius cohortem impuram cum illius exercitu invicto, huius libidines cum illius continentia.*

47. Cic. *Verr.* 2.4.115: *Ab illo qui cepit conditas, ab hoc qui constitutas accepit captas dicetis Syracusas.*

48. The passage refers repeatedly to Marcellus (2.4.115, 116, 120, 121, 122, 123, 130, 131) and marks its return to him explicitly (2.4.120, 131).

49. The most well-known instance of this phenomenon in Cicero is the proso-popoeia at *Cael.* 33–34 (Appius Claudius Caecus rebukes Clodia). Roller ("Exemplarity," 37) points out this and other instances of the "ideologically potent inversion, where the exemplary hero is revivified to sit in judgment on posterity," including the epitaph of Cn. Cornelius Scipio Barbatus (*CIL* 12.6–7) and Polyb. 6.53–54, on the Roman aristo-cratic funeral. "Posterity" here includes Marcellus's own family, for in addressing both the *iudex*, C. Marcellus, and the witness for the defense, M. Marcellus Aserninus, Cicero addresses potentially "good" and "bad" members of the clan and raises up the ghost of the great Marcellus to judge them.

50. Livy 25.24.11. Marincola ("Marcellus at Syracuse," 227) points out, rightly, that this cannot be an "Alexandrian footnote" flagging an allusion to a specific predecessor, because historians from Herodotus onward recorded variant traditions. Still, to record a variant is to place one's own version in the context of tradition and thus remind readers that there is a tradition.

51. For example, he recalls the battles of the Sicilian Expedition. See Rossi, "Tears of Marcellus"; Jaeger, "Livy and the Fall of Syracuse," 230–34. Marincola ("Marcellus at Syracuse," 228) puts it well: "Livy is reflecting on the movement of history and on his own role as a chronicler of that movement."

52. In Livy, this speech characterizes, indirectly, both Marcellus, whom it sways, and those superb rhetoricians the Sicilians, who know precisely what words will capture the imagination of an ambitious Roman aristocrat: *gloria, memorandum, triumphi titulus, fama, posteris spectaculum, tropaea, clientela, nomen Marcellorum.* On the rich resources of Roman aristocratic memory, see Hölkeskamp, "*Exempla* und *mos maiorum,*" 305–8.

53. To rephrase the problem as a question, does Livy speak here as a moralizing his-torian making a comment on the event, as a reader of sources making a comment on them, or as both? On the voices that articulate exempla in Livy, see J. D. Chaplin, *Livy's Exemplary History* (Oxford: Oxford University Press, 2000), 50–72.

54. See F. Münzer, "Fabius Pictor" *RE* 6.2 (1909): 1836–41; fragments in *HRR* I. H. Beck and U. Walter (*Die frühen römischen Historiker,* vol. 1, *Von Fabius Pictor bis Cn. Gellius* [Darmstadt: Wissenschaftliche Buchgesellschaft, 2001], 54–136) offer a more re-cent edition of the fragments, with translation, commentary, and extensive bibliogra-phy. See also Frier, *Libri Annales Pontificum Maximorum,* 227–53. Frier (237–38) suggests that Pictor may have begun writing after the fall of Syracuse, "within this climate of cau-tious hope," and thinks that either death or old age prevented him from describing the events of 213 and after. Ernst Badian ("The Early Historians," in *Latin Historians,* ed. T. A. Dorey [New York: Basic Books, 1966], 1–38) thinks it more likely that he wrote after the war ended.

55. Gruen (*Studies in Greek Culture,* 94) places emphasis on the play as a pointer to collective success. This would make Marcellus all the more the representative Roman.

56. Flower (*Ancestor Masks and Aristocratic Power,* 91–127) points out that funerals offered opportunities for the dramatic representation of the history of the gens; and T. P. Wiseman has argued vigorously for the early influence of drama on the shaping of Roman historical narrative. In *Roman Drama and Roman History* (Exeter: University of Exeter Press, 1998), 31, Wiseman speaks of the Tullia story (Livy 1.46–48) as one that Livy may have known from the stage, and he compares the ways in which various authors

handle the preliminary murders of Tullia's sister and Tarquin's wife. Wiseman notes that none of the authors say how the murder was done, and he observes that it appears in "a subordinate clause in Livy, a throwaway sentence in Dionysius, an ablative absolute in Ovid." He points out that "it is an obvious possibility that the authors' narrative emphasis reproduced scenes on stage—soliloquy, dialogue, dramatic conflict—and what happened off stage was out of sight and out of mind." In *Historiography and Imagination: Eight Essays on Roman Culture* (Exeter: University of Exeter Press, 1994), 18, Wiseman points out, "I am not, of course, suggesting that Livy or Dionysius took such episodes directly from dramatic sources; but I do suggest that much of the traditional material in early Rome which they found in their second- and first-century BC predecessors had been processed for the stage long before it ever appeared in the pages of any historian." A dramatic "Fall of Syracuse"—featuring Marcellus weeping and soliloquizing over the city in one scene, Archimedes protecting his diagrams with broad gestures and the bold words attributed to him as the unknown soldier draws his sword (end of scene), Marcellus in the next scene receiving news of the death and then displaying grief and ordering up a good (Roman) burial—makes for good drama and even better public relations (all Romans should have the Syracusan artwork, because the Roman Marcellus could appreciate Syracusan genius). Could this narrative have been the center of a trilogy (a "Clastidium," a "Fall of Syracuse," a "Death of Marcellus")? In "The Tragic History of Marcellus," Carawan in Livy's third decade, shows how Livy uses the structure of tragedy to organize Marcellus's campaigns (Nola, Syracuse, death in ambush). McDonnell (*Roman Manliness*, 228–35) argues that displaying the spoils of Syracuse was Marcellus's innovative way of mustering popular support. Dramatic presentation of the siege and sack could have contributed to such an effort.

57. On the Greek tradition, see Frier, *Libri Annales Pontificum Maximorum*, 280–84.

58. E.g., Livy 27.28.1.

59. Livy 25.31.11: *hoc maxime modo Syracusae captae.*

60. Livy (24.34.2) also calls Archimedes *unicus spectator caeli siderumque.*

61. Cf. *Aen.* 5.814–15 (Neptune predicting the death of Palinurus): *unus erit tantum amissum quem gurgite quaeres; / unum pro multis dabitur caput* (one only will there be lost, whom you will seek in the whirlpool; one head will be given for many).

62. See David Konstan, "Narrative and Ideology in Livy: Book I," *CA* 5 (1986): 198–215.

63. Cf. Livy 1. 7. 2: *ludibrio fratris;* 24.34.2: *ipse perlevi momento ludificaretur;* 24.34.12: *ita maritime oppugnatio elusa est;* 24.34.16: *omnis conatus ludibrio esset.*

64. Livy's "Death of Archimedes" (25.31.9–11) and his "Death of Remus" (1.7.2) share a similar presentation: a death motivated by anger (*ira / irato*); an account of death as tradition (*fama est / proditum memoriae*); an indirect statement leading to a strikingly delayed participle, *interfectum* (R. M. Ogilvie, *A Commentary on Livy, Books 1–5* [Oxford: Oxford University Press, 1965], ad loc.); and a final summary of the city's fate and the victor's reward: *ita solus potitus imperio Romulus; condita urbs conditoris nomine appellata* (thus Romulus alone gained the rule, and the city, now founded, was called by the name of its founder) / *hoc maxime modo Syracusae captae; in quibus praedae tantum fuit, quantum vix capta Carthagine tum fuisset* (and this, for the most part, was how Syracuse was captured, in which city there was as much booty as there would scarcely have been at the time if Carthage had been captured).

65. Cic. *Verr.* 2.4.115.

66. Ennius, that great filter of republican exempla, may have told the story briefly.

See H. W. Litchfield, "National *Exempla Virtutis* in Roman Literature," *HSCP* 24 (1914): 1–71.

67. Text from J. Delz, ed., *Silius Italicus, Punica* (Stuttgart: Teubner, 1987).

68. See A. Pomeroy "Silius Italicus as '*Doctus Poeta*,'" *Ramus* 18, nos. 1–2 (1989): 119–39.

69. Note that in addition to Archimedes' *industria* (Valerius) and the soldier's *imprudentia* (Pliny), we now have a third abstract killer, *ruina*, which belongs, apparently, to nobody.

70. Compare Livy, where the following events intervene between Marcellus's tears (25.24.11) and the death of Archimedes (25.31.9): attempts at negotiations, visits to Marcellus by delegations from Neapolis and Tyche, Marcellus's agreement to harm no free man, the Roman plunder of Neapolis and Tyche, the investment of Achradina, the plague, the near arrival of Carthaginian reinforcements, more surrender talks, a night assault on Achradina, and Syracuse's final surrender. Note also that Silius (14.580–640) describes the plague at length before describing Marcellus's tears.

71. Cf. *Aen.* 6.30–31: *tu quoque magnam / partem opere in tanto, sineret dolor, Icare, haberes* (you also, Icarus, would have a great part in such a work, if grief permitted); 7.1–4: *Tu quoque litoribus nostris, Aeneia nutrix / aeternam moriens famam, Caieta, dedisti / et nunc servat honos sedem tuus; ossaque nomen / Hesperia in magna, si qua est ea gloria, signat* (You also, Caieta, nurse of Aeneas, gave fame to our shores, when you died; and even now your honor preserves your resting place, and if it is any glory, your bones mark your name in great Hesperia). Parallels occur in Silius at 3.287, 4.635, 6.537, and 15.451. See Fr. Spaltenstein, *Commentaire des Punica de Silius Italicus*, 2 vols. (Geneva: Droz, 1986–90), ad loc.

72. *Aen.* 10.793: *non equidem nec te, iuvenis, memorande silebo* (nor shall I keep silent [your deeds] or you, yourself , young man, who are be remembered); 10.788–89: *ingemuit clari graviter genitoris amore / ut vidit Lausus: lacrimaeque per ora volutae* (when Lausus saw this, he groaned heavily with love for his illustrious father, and tears rolled down his cheeks); 10.821: *ingemuit miserans graviter, dextramque tetendit* (He groaned heavily in pity and extended his right hand).

73. Cf. Aeneas's response to the depictions of Troy on the temple of Juno (*Aen.* 1.462): *sunt lacrimae rerum et mentem mortalia tangunt* (there are tears for things, and the mortal condition touches the mind). Virgil links the narrator's obligation to commemorate more emphatically to *memorande* than does Silius. See *Aen.* 10.793; *Georg.* 3.1–2.

74. On the figure of Marcellus in the *Punica*, see Frederick M. Ahl, M. A. Davis, and Arthur Pomeroy, "Silius Italicus," *ANRW* II.32.4 (1986): 2536–42, pointing out the connection to *Aeneid* 6. Why do we get there via Archimedes? Andrew Feldherr has suggested to me that the passage could be a "window reference" to an earlier epic version of the Archimedes story.

75. The perfect *fuit* in *sic parcere victis pro praeda fuit* is telling: such mercy is a thing of the past. Ahl, Davis, and Pomeroy ("Silius Italicus," 2539) explain: "With the end of *Punica* 14, we reach the end of Roman mercy to the defeated foe. As the young leaders of newer generations assume command, Carthage will be treated by the conquering Romans with a ferocity that outdoes even that of Claudius Nero." On the museumlike quality of the historical past in the *Punica*, see C. Santini, *Silius Italicus and His View of the Past* (Amsterdam: J. C. Gieben, 1991), 9.

76. *Aen.* 6.854–56: *Sic pater Anchises, atque haec mirantibus addit: / 'aspice, ut insignis*

spoliis Marcellus opimis / ingreditur victorque viros supereminet omnis' (Thus spoke Father Anchises, and as they marveled, he added these words: "see how Marcellus goes along, conspicuous with his rich spoils, and as victor towers above all men").

77. Ahl, Davis, and Pomeroy ("Silius Italicus," 2499) point out a good example of such a rereading of *Aeneid* 6 at *Punica* 7.106–7, where Hannibal asks, *en ubi nunc Gracchi atque ubi nunc sunt fulmina gentis Scipiadae?* Ahl, Davis, and Pomeroy point out that although Hannibal is thinking "not of those probably uppermost in Anchises' mind, but of the men he has already defeated (instead of those who will defeat him)," he also anticipates/recalls *Aen.* 6.842–43.

78. Ahl, Davis, and Pomeroy, "Silius Italicus," 2538.

79. Silius's very omission of Archimedes' name may help emphasize the absence of a tomb, by drawing attention to the fact that the name does not fit the meter traditionally associated with epitaphs. The necessity of including names unsuited to elegiacs was a frequent cause of metrical eccentricities. See R. Kassel, *"Quod Versu Dicere Non Est,"* *ZPE* 19 (1975): 211.

80. On the way in which Silius plays with endings, see Philip Hardie, "Closure in Latin Epic," in *Classical Closure: Reading the End in Greek and Latin Literature,* ed. D. H. Roberts, Fr. Dunn, and D. Fowler (Princeton: Princeton University Press, 1997), 139–62.

81. *OLD,* s.v. *fero* 36.a.

82. My thanks to William Dominik at the University of Otago for drawing my attention to how Silius undermines the representation of Roman virtue. See Dominik's "Hannibal at the Gates."

83. My thanks to Robert Sklenar at the University of Tennessee for drawing my attention to this point.

CHAPTER 5

1. The sources, for example, generally overlook the fact that the very weapons used against Syracuse were Greek in origin. Moschus, cited by Athenaeus (14.634), said that the Romans' new weapon was invented by Heracleides of Tarentum (Μόσχος δ' ἐν πρώτῳ Μηχανικῶν Ῥωμαϊκὸν εἶναι λέγει τὸ μηχάνημα καὶ Ἡρακλείδην τὸν Ταραντῖνον εὑρεῖν αὐτοῦ τὸ εἶδος). Athenaeus refers to this engine, to Biton on its form and construction, and to Andreas of Panormus for its being called a *sambuca* as a result of its harplike appearance. This seems to be a case of Greek one-upmanship. On the weapons at Syracuse, see Walbank, *Historical Commentary,* 2:72–74; J. G. Landels, "Ship-Shape and Sambuca-Fashion," *JHS* 86 (1966): 69–77; J. G. Landels, *Engineering in the Ancient World* (Berkeley and Los Angeles: University of California Press, 1978), 95, 97, 99–132; K. D. White, *Greek and Roman Technology* (Ithaca: Cornell University Press, 1984), 47–48. On the fortifications, see A. W. Lawrence, "Archimedes and the Design of Euryalus Fort," *JHS* 66 (1946): 99–107.

2. For example, Plutarch (*Marc.* 8.6) says that after the battle at Clastidium, Rome sent spoils to Hieron, who was its friend and ally (φίλον ὄντα καὶ σύμμαχον). On Rome's alliance with Syracuse, see Eckstein, *Senate and General,* 102–34. McDonnell (*Roman Manliness,* 231), noting this alliance and pointing out that Marcellus fought in Sicily as a young man, suggests that the court of Hieron II influenced Marcellus's ideas about the political uses of public art. If Marcellus visited Syracuse, he could have met Archimedes.

3. Particularly useful for the overall conception of this chapter is Michel Authier's "Archimedes: The Scientist's Canon." Authier notes additional signs of the neatening up of the past: he points out (137–38) that Syracuse had exported war engines from the fourth century and that, in consequence, the vigor of its defense could scarcely have come as a surprise. Feeney ("Beginnings of a Literature in Latin," 239, citing Momigliano, *Alien Wisdom*, 38) points out the interest of pro-Roman Greeks, like Polybius, in preserving "their image of the noble untainted Roman."

4. On Plutarch, see C. P. Jones, *Plutarch and Rome* (Oxford: Oxford University Press, 1971); T. Duff, *Plutarch's Lives: Exploring Virtue and Vice* (Oxford: Oxford University Press, 1999); Pelling, *Plutarch and History*; B. Scardigli, ed. *Essays on Plutarch's Lives* (Oxford: Oxford University Press, 1995).

5. On this digression, see A. Wardman, "Plutarch's Methods in the *Lives*," *CQ* n.s. 21 no. 1 (1971): 254–61; A. Georgiadou, "Bias and Character-Portrayal in Plutarch's *Lives* of Pelopidas and Marcellus," *ANRW* II.33.6 (1992): 4222–57; A. Georgiadou, "The Corruption of Geometry and the Problem of Two Mean Proportionals," in *Plutarco e le scienze (Atti del IV convegno plutarcheo, Genova-Bocca di Magra, 22–25 Aprile, 1991),* ed. I. Gallo (Genoa: Sage, 1992), 147–64. I agree with Georgiadou ("Corruption of Geometry," 147) that, pace Wardman ("Plutarch's Methods," 257–58), this digression is relevant to the life of Marcellus, but I read it very differently.

6. Cf. Pappas of Alexandria (*Mathematical Collection* 8.9): δός μοί πoῦ στῶ, καὶ κινῶ τὴν γῆν. Tzetzes (*Chil.* 2.144) draws attention to Archimedes' dialect: Ἔλεγε δὲ καὶ δωριστὶ φωνῇ Συρακουσίας Πᾶ βῶ καὶ χαριστίωνι τὰν γᾶν κινήσω πᾶσαν (And he said in the Doric speech of Syracuse, give me a place to stand, and I shall move the entire earth with a triple pulley). Paul Ver Eecke ("Note sur une interprétation erronée d'une sentence d' Archimede," *L'Antiquité classique* 24 [1955]: 132–33) points out that this expression appears nowhere in Archimedes' works.

7. For other cross-cultural foils, see Plut. *Pyrrhus* 20.1–21.5, in which the poor but virtuous Roman Fabricius stands in contrast to the king; for discussion, see S. C. R. Swain, "Hellenic Culture and the Roman Heroes of Plutarch," in Scardigli, *Essays on Plutarch's Lives*, 250–51. Pelling (*Plutarch and History*, 106) says that any account of the lives must "bring out their *versatility.*" This includes finding room for "Lives which break away from the constrictions of a single man's Life, as *Antony* moves its interest to Cleopatra, or as *Brutus* often divides its interest between Brutus and Cassius." For more on the effects of such intrusions, especially accounts of the deaths of other characters, see Pelling, *Plutarch and History*, 365–86.

8. Swain ("Hellenic Culture," 239, 257) observes that the "emphasis on Hellenism may be Plutarch's development of Marcellus' description of Archimedes, in which he alludes to the culture of the symposion (17.2)," and that Plutarch "may have developed the idea of Marcellus' παιδεία from his famous remark about Archimedes." I think that Swain is correct, and I will explore the effects of this development.

9. Tzetzes *Chil.* 2.149–52: "Dio and Diodorus write the story (γράφει τὴν ἱστορίαν); and many with them recall Archimedes: first of all, Anthemius, the paradoxographer, and Hero and Philo and Pappus, and every writer on mechanics (πᾶς μηχανογράφος)."

10. Ὁ Ἀρχιμήδης ὁ σοφὸς μηχανητὴς ἐκεῖνος, / τῷ γένει Συρακούσιος ἦν, γέρων γεωμέτρης, / χρόνους τε ἑβδομήκοντα καὶ πέντε παρελαύνων, / ὅστις εἰργάσατο πολλὰς μηχανικὰς δυνάμεις, / καὶ τῇ τρισπάστῳ μηχανῇ χειρὶ λαιᾷ καὶ μόνῃ /

πεντεμυριομέδιμνον καθείλκυσεν ὁλκάδα. / Καὶ τοῦ Μαρκέλλου στρατηγοῦ ποτέ δε τῶν Ῥωμαίων / τῇ Συρακύσῃ κατὰ γῆν προσβάλλοντος καὶ πόντον, / τινὰς μὲν πρῶτον μηχαναῖς ἀνείλκυσεν ὁλκάδας, / καὶ πρὸς τὸ Συρακούσιον τεῖχος μετεωρίσας / αὐτάνδρους πάλιν τῷ βυθῷ κατέπεμπεν ἀθρόως (text from P. A. M. Leone, ed., *Ioannis Tzetzae, Historiae* [Naples: Università degli Studi di Napoli, 1968]).

11. Authier, "Archimedes"; see especially 124–25. Authier's exploration of the relation between science and power begins with the story of Daedalus, whom he sees as a forerunner of Archimedes, who is, in turn, the original of the Soviet physicist and Nobel Prize winner Sakharov, the atomist Oppenheimer, and the Nobelist Frédéric Joliot-Curie. For further sociopolitical discussion of this story, see Bruno Latour, *We Have Never Been Modern*, trans. Catherine Porter (Cambridge, MA: Harvard University Press, 1993), 109–11.

12. P. Culham, "Plutarch and the Roman Siege of Syracuse: The Primacy of Science over Technology," in Gallo, *Plutarco e le scienze*, 179–97.

13. Hieron asks for "something large moved by a small force" (τι τῶν μεγάλων κινούμενον ὑπὸ σμικρᾶς δυνάμεως, *Marc.* 14.8). It is a miniature both in size and in the time required for the experiment: Authier ("Archimedes," 137) notes that the eighteenth-century Scottish philosopher Adam Ferguson calculated that "it would take a man pushing against the earth at the end of a lever and moving at the speed of a cannonball" 44,963,540,000,000 years to move the earth one inch.

14. Authier ("Archimedes," 135–36) points out the thematic parallel, noted by Silius Italicus (*Punica* 14.349–51) between counting the grains of sand in the universe and one man moving the earth: the first measures the greatest in terms of the smallest; the second suggests that the heaviest can be moved by the lightest. Each involves an extreme change in scale. Netz (*Ludic Proof*, chap. 1, "The Carnival of Calculation") shows how the Hellenistic obsession with size and vast numbers plays into Archimedes' presentation of his proofs.

15. *Punica* 14.351–52: *puppis etiam constructaque saxa feminea traxisse ferunt contra ardua dextra.* On *feminea dextra* as connoting feebleness, see Spaltenstein, *Commentaire des Punica*, 2:312.

16. Plutarch (*Demetrius* 43.5–6) describes this ship and says that it ended up being only for show, since it was difficult to move.

17. R. Webb, "Picturing the Past: Uses of Ekphrasis in the *Deipnosophistae* and Other Works of the Second Sophistic," in *Athenaeus and His World: Reading Greek Culture in the Roman Empire*, ed. David Braund and John Wilkins (Exeter: University of Exeter Press, 2000), 218–26.

18. Webb, "Picturing the Past," 225.

19. Lucian *Nav.* 5–6.

20. Heron of Alexandria wrote three books on mechanics, the first of which presented ways of moving great weights with little effort. See A. G. Drachmann, *The Mechanical Technology of Greek and Roman Antiquity* (Copenhagen: Munksgaard, 1963), 19–140.

21. Polyb. 8.3.3. See Walbank, *Historical Commentary*, 2:72–74. . The text is from Th. Büttner-Wobst, ed., *Polybius, Historiae*, vol. 2 (1889; reprint, Leipzig: Teubner, 1985).

22. Authier ("Archimedes," 158) identifies "the effect of the minimum on the maximum" as "the fundamental framework of Archimedes' thought." The first chapter of Netz's *Ludic Proof* elaborates this idea.

23. Polyb. 8.4.2–11; see also 5.37.10 (σαμβυκίστρια). For how it may have looked, see Landels, "Ship-Shape"; Walbank, *Historical Commentary*, 2:72. See also Paul Keyser and Georgia Irby-Massie, "Science, Medicine, and Technology," in *The Cambridge Companion to the Hellenistic World*, ed. G. R. Bugh (Cambridge: Cambridge University Press, 2006), 241–64, with a diagram of the *sambuca* on 260.

24. He gives it roughly three times the space (thirty-five Teubner lines for the *sambuca*; eleven for the Hand).

25. The ratio is twenty-seven words to forty-nine. See Livy 24.34.6–7 for the description of the *sambuca*, 24.34.10 for the *ferrea manus*.

26. This translation is modified from Perrin, *Plutarch's Lives*.

27. It is not clear from either Plutarch or Polybius precisely how the machine worked. Polybius's description is at 8.6.1–4; see Walbank, *Historical Commentary*, 2:75–77. On Plutarch's lack of interest in the technology, see Culham, "Plutarch and the Roman Siege of Syracuse," 180–81.

28. Spaltenstein, *Commentaire*, 2, *ad* 14.316.

29. Tzetzes has not received much respect as a poet; see N. G. Wilson, *Scholars of Byzantium* (Baltimore: Johns Hopkins University Press, 1983), 190–96. This passage, however, has an interesting, almost antiphonal structure, in which Marcellus's separate attacks, each from a different distance, are met with the proper defense.

30. The metaphor captured the imagination of the Renaissance painter Giulio Parigi, whose decoration of the Hall of Mathematicians in the Uffizi Gallery shows a machine that really looks like an enormous hand, to striking effect. The Hand dominates the foreground of the painting, while the *sambuca* is only a small image in the distant background. See P. Galluzzi, *Archimede e la Storia delle Matematiche nella Galleria degli Uffizi* (Syracuse: A. Lombardi, 1989), with plates.

31. As Authier observes ("Archimedes," 151–52), Plutarch's account omits any reference to the careful coordination of this defense. Authier also argues that Plutarch passes over Marcellus's organization of the blockade in order to explain the fall of Syracuse by means of the "three enemies of science": time, lies, and superstitious religion. In "We Have Never Been Modern," Latour, too, points out how Plutarch elides organization and politics.

32. Σαμβύκη could mean the harpist as well as the instrument. On ὥσπερ ἐκσπόνδους, see Walbank, *Historical Commentary*, 2:77.

33. On this hostility, see chap. 4.

34. Livy 24.33.9: *non diffidebant uastam disiectamque spatio urbem parte aliqua se invasuros.*

35. Livy 24.34.16: *ita consilio habito, quando omnis conatus ludibrio esset, absistere oppugnatione atque obsidendo tantum arcere terra marique commeatibus hostem placuit.* These collective expectations and decisions also appear in Polybius; what is striking is Livy's omission of Marcellus's point of view.

36. My thanks to Andrew Feldherr for pointing out this effect.

37. On *ludibrium*, see P. Plass, *Wit and the Writing of History* (Madison: University of Wisconsin Press, 1988), 15–25. Plass observes that "its element of jest is often a vicious sport or ugly theater of the absurd" (16).

38. The Teubner text (K. Ziegler and H. Gärtner, eds. *Plutarchis Vitae Parallelae*. vol. 4. Leipzig: Teubner, 1980) prefers πρὸς τὴν θάλασσαν to ἐκ τῆς θαλάσσης, which is an early correction based on Polybius. The phrase "in respect to the sea" is stilted, but it is

not clear how else to render the prepositional phrase, unless by "toward the sea." This passage has several textual problems (see the apparatus in Ziegler and Gärtner), which have been complicated by the parallels from Polybius and Athenaeus, but the general image is clear: that of the Hand holding the ships, as if they were κύαθοι.

39. On Plutarch's jokes, see T. Reekmans, "Verbal Humor in Plutarch and Suetonius' Lives," *Ancient Society* 23 (1992): 189–232. Cicero, too, made jokes that depended on scale, about his brother Quintus and son-in-law Lentulus, both men of short stature. Seeing a statue of Lentulus girded with a great sword, he asked, "Who tied my son-in-law to a sword?" When he had come into the province that his brother was governing and saw a most imposing bust of his brother with his shield, he said, "My brother halved is taller than his whole" (Macrobius 2.3.3–4).

40. T. Nagel, *Mortal Questions* (Cambridge and New York: Cambridge University Press, 1979), 11–12.

41. Nagel, *Mortal Questions*, 15.

42. On Marcellus's warlike character, see *Marc.* 1.2; on his courage, see 2.1. When Plutarch reports Marcellus's prayer vowing the *spolia opima* after the battle of Clastidium, he has Marcellus say that he killed the enemy "with his own hand" (ἰδίᾳ χειρὶ).

43. Hand and *sambuca* do not even meet directly in Plutarch's narrative. He tells of stones of ten talents' weight being fired at and shattering the *sambuca* (*Marc.* 15.3–4).

44. E.g., Swain, "Hellenic Culture," 257.

45. Plut. *Marc.* 1.2–3. See Plutarch's *Life of Fabius Maximas* 19.1–4 for a similar description of Marcellus's warlike nature. The *Life of Marcellus* is unique in emphasizing its subject's duality so strongly at the start. This lack of leisure is part of the construct of the Roman character. Dench (*From Barbarians to New Men*, 30) explains: "By the late Republic Romans were themselves hard at work constructing the 'true' and ancient Roman character, busy 'doers' permanently involved in military matters, without the time or inclination to turn to softer, lazier, and altogether more decadent things such as sitting around and talking or writing books. Such pursuits came later, in an inferior social and moral climate." Marcellus's inability to pursue his Hellenic interests calls to mind not only Sallust's *Cat.* 1–13 (cited by Dench) but also, ironically, Plutarch's own autobiographical statement (*Dem.* 2) that, while he was at Rome, his business and philosophical teaching prevented him from completely mastering Latin. He had become a "doer" too, even in the matter of philosophy, serious stuff for a Greek.

46. Plut. *Marc.* 1.5: κατ' εὐγένειαν καὶ ἀρετήν.

47. Plut. *Marc.* 14.7: συγγενὴς ὢν καὶ φίλος.

48. At *Marc.* 30.5, Plutarch says that Cornelius Nepos and Valerius Maximus report Marcellus was left unburied but that Livy and Augustus report his urn was brought to his son and honored with a splendid funeral.

49. This is a statue of Marcellus at Lindos (Plut. *Marc.* 30.7–8).

50. Plut. *Marc.* 1.1: Marcellus is strong in body (τῷ δὲ σώματι ῥωμαλέος); 2.1: he saves his brother in battle; 30.1: Hannibal stood over Marcellus's corpse, "eyeing its strength and form" (τήν τε ῥώμην τοῦ σώματος καταμαθὼν καὶ τὸ εἶδος).

51. A. Wardman, *Plutarch's Lives* (Berkeley and Los Angeles: University of California Press, 1974), 8.

52. For other examples of this traditional opposition between παίζω and σπουδάζω, see LSJ, s.v. παίζω, citing Xen. *Mem.* 4.1.1; Xen. *Cyr.* 8.3.47; Plato, *Euthydemus* 283b.

53. That the joke played on the customs of the symposium would have made it an especially attractive candidate for expansion. See Wardman, "Plutarch's Methods." Wardman (256) observes of Plutarch's comment about jokes in *Alex.* 1.2: "Plutarch is not expressing an interest in παιδιά for its own sake, but has, I think, derived the notion from Plato, who sees in παιδιά, as exemplified in symposia, a safe and certain way of discovering the φύσεις of those present."

54. In "Bias and Character-Portrayal," Georgiadou identifies bias in favor of Pelopidas.

55. *Marc.* 18.1: Ἀρχιμήδης μὲν οὖν τοιοῦτος γενόμενος ἀήττητον ἑαυτόν τε καὶ τὴν πόλιν, ὅσον ἐφ' ἑαυτῷ, διεφύλαξε.

56. Marcellus's fatal inability to control his ambition was due to insufficient *paideia*, according to Swain ("Hellenic Culture," 240, 254–59). Pelling (*Plutarch and History,* 285–86) makes this point too.

57. S. C. R. Swain, *Hellenism and Empire: Language, Classicism, and Power in the Greek World, AD 50–250* (Oxford: Oxford University Press, 1996), 139–40.

58. Of course, there were also other things that turned out contrary to Roman expectations, such as fighting the war in Italy, not Spain (Polyb. 3.15.13, 3.16.5–6).

59. See Craige B. Champion, *Cultural Politics in Polybius's "Histories"* (Berkeley, Los Angeles, and London: University of California Press, 2004), app. C, on *logismos* in Polybius's *Histories*. Champion (255) maintains that "a Polybian image of Romans (especially in their earlier history) is that of quasi Hellenes who possess the quintessential Hellenic virtue of reasoning power, or *logismos*."

60. Champion (*Cultural Politics,* 235) says: "In our search for the Romans on a Polybian Hellenic-barbarian continuum, we have found that the Romans do not occupy a fixed position; rather they slide between the poles of Hellenism and barbarism." The failure of imagination at this point may also reflect Polybius's hostile portrait of Marcellus.

61. Netz (*Ludic Proof,* chap. 1, "The Carnival of Calculation") shows how the Hellenistic obsession with size and vast numbers plays into Archimedes' presentation of his proofs.

62. I am grateful to Reviel Netz for pointing this out to me (see LSJ, s.v. δύναμις V).

63. In *Ludic Proof* (chap. 1), Netz argues that the mathematical universe set up by Hellenistic treatises of calculation is one of loosened control and suspended reason; this is in contrast to the perfect control and pure abstraction of Greek geometry as made known by Plato. Polybius's scene is almost a demonstration of old math meeting new math.

64. Richard Martin ("Seven Sages," 113) singles out three criteria that mark the wise man: he is a poet, he is involved in politics, and he is a performer. As to the first, Archimedes' *Cattle Problem* (how to count the cattle of the Sun), as we have it, is in elegiac couplets, with the prose heading, "Which Archimedes solved in epigrams and which he communicated to students of such matters in a letter to Eratosthenes of Cyrene" (Bulmer-Thomas, ed., *Greek Mathematical Works,* vol. 2). Although the epigram, as we have it, is probably not the work of Archimedes himself, still, an ancient tradition connected him with poetry. Cicero, moreover, associates Archimedes with a poem that, if not gnomic, at least contains a universal truth and is mnemonically potent. As to the second criterion, involvement in politics, although we have no record of explicitly political acts or of any laws or treaties (besides treatises on the laws of the universe) associated with Archimedes, as we do for other wise men, nevertheless, several sources—Polybius, Vitru-

vius, Livy, Plutarch, and Proclus—link him to Hieron II and his son, Gelon. Archimedes fits the third criterion particularly well: "clever demonstrations" also mark the wise man. Martin (116) points out, "What might be read in later time as 'engineering' feats can just as easily be in the realm of such clever demonstrations." He cites as an example Thales' diverting the Halys River so that Croesus could cross it without a bridge (Hdt. 1.75). Archimedes' mechanical demonstrations, including moving the big ship and defending Syracuse, fit nicely into this category. Moreover, Martin notes (116) that "the performing sages often have a high-status audience, whether Greek or barbarian." Archimedes has both, the Greek Hieron (Gelon as well, in Proclus) and the "barbarian" Marcellus.

65. On vision and display in Polybius, the fundamental work is J. Davidson, "The Gaze in Polybius' *Histories*," *JRS* 81 (1991): 10–24. See also A. D. Walker, "*Enargeia* and the Spectator in Greek Historiography," *TAPA* 123 (1993): 353–77.

66. This fatal flaw appears in other Roman lives. Swain (*Hellenism and Empire*, 143) points out that "Cicero's surrender to a love of glory that is at variance with philosophical training is a major theme of his life." Swain (185–86) summarizes Plutarch's view of Rome: "While both Greeks and Romans were capable of producing men whose lives were suitable for imitation in our own, there was a lurking suspicion that Romans lacked proper, Greek culture, which was the only path to philosophical happiness."

67. The quoted phrase is from Dench, *From Barbarians to New Men*, 30.

CODA TO PART TWO

1. An Egyptian from Alexandria, Claudius Claudianus came to Rome as a young man in 394 CE and became the court poet to the emperor Honorius. On his life and work in general, see A. Cameron, *Claudian: Poetry and Propaganda at the Court of Honorius*. Oxford: Oxford University Press, 1970. The text is from J. B. Hall, ed., *Claudianus, Carmina* (Leipzig: Teubner, 1985).

2. See Cic. *Rep.* 1.21; *Nat. D.* 2.88; *Tusc.* 1.63. See also Quacquarelli, "La fortuna di Archimede," which thoroughly discusses the late antique sources.

3. Gee, *Ovid, Aratus, and Augustus*, 96–100.

4. Sextus Empiricus *Adversus grammaticos* (Against the Professors) 115; Lactantius *Div. inst.* 2.5.18.

5. Martianus Capella 6.583; Cassiodorus *Variorum* 1.45; Cic. *Nat. D.* 2.88: *quis in illa barbaria dubitet quin ea sphaera sit perfecta ratione?* (who in that barbarian realm would doubt that this sphere was fashioned by means of reason?). The point about machines impressing barbarians is made by D. Shanzer in "Two Clocks and a Wedding," *Romano-barbarica* 14 (1996–97): 225–58.

6. *Claudianus, Carmina*, J. B. Hall, ed., Leipzig: Teubner, 1985.

7. On Claudian's scientific knowledge and representations of the natural world, see Cameron, *Claudian*, 343–48; P. Fargues, *Claudien: Études sur sa poésie et son temps* (Paris: Hachette, 1933), 311–20. On Claudian's frequent references to astronomy and astrology, see Fargues, *Claudien*, 174–81. Compare to *Carmina minora* 51 the expression of universal order at the beginning of Claudian's *In Rufinum* (1.4–11).

8. M. J. Roberts, *The Jeweled Style: Poetry and Poetics in Late Antiquity* (Ithaca: Cornell University Press, 1989), 19–37.

9. Note the following adjectives in *Carmina minora* 34: *inclusus, interior, arcano*.

10. Two Greek epigrams, *Anth. Pal.* 9.753–54, also on crystal balls, are attributed to

Claudian. See Cameron, *Claudian,* 12–14. For recent discussion of minerals in late antique poetry, see D. Petrain, "Gems, Metapoetics, and Value: Greek and Roman Responses to a Third-Century Discourse on Precious Stones," *TAPA* 135 (2005): 329–57.

11. The poem appears in general to avoid the interest in material adornment that typifies what Roberts calls the "jeweled style" (in Roberts, *Jeweled Style*).

CHAPTER 6

1. The Latin texts of Petrarch are from the national edition: G. Martellotti, ed., *De Viris Illustribus,* vol. 2 of *Edizione nazionale delle opere di Francesco Petrarca* (Florence: Sansoni, 1964); G. Billanovich, ed., *Rerum Memorandarum Libri,* vol. 5 of *Edizione nazionale delle opere di Francesco Petrarca* (Florence: Sansoni, 1943). For general background to this chapter, see Nicholas Mann, "The Origins of Humanism," and M. D. Reeve, "Classical Scholarship," in *The Cambridge Companion to Renaissance Humanism,* ed. J. Kraye (Cambridge: Cambridge University Press, 1994), 1–19, 20–46. For overviews of Petrarch's life, see E. H. Wilkins, *Life of Petrarch* (Chicago and London: University of Chicago Press, 1961); Kenelm Foster, *Petrarch: Poet and Humanist* (Edinburgh: Edinburgh University Press, 1984); Nicholas Mann, *Petrarch* (Oxford and New York: Oxford University Press, 1984). Petrarch also uses Archimedes as an exemplum in *De remediis* 1.99.

2. The *De viris illustribus,* never completed, comprises twenty-three ancient lives, written 1338/9–43. Petrarch later added twelve pre-Romulan lives (1351–52) and a preface (1351–53). See Martellotti, *De Viris,* ix–x; Foster, *Petrarch,* 148, 181–82.

3. The *RM* was taking shape by December 1343. Left off in February 1345, it languished like the *De viris illustribus.* See Billanovich, *Rerum Memorandarum,* cvi–cviii. According to Wilkins (*Life of Petrarch,* 48–49), Petrarch completed book 1 of the *RM* before leaving Provence for a trip to Naples in 1343.

4. For Val. Max. 8.7.7, see chap. 4 of the present study; the other authors are Cicero, Livy, and (in the *Marcellus*) Firmicus Maternus. The extraordinary popularity of Valerius Maximus in the Middle Ages largely determined the shape of the biographical tradition handed down to Latin readers in the West. The texts, collected by Clagett, that preserve the medieval Latin biographies of Archimedes almost invariably quote or paraphrase Val. Max. 8.7.7 (although some also show the influence of Orosius and Lactantius). See Clagett, *Archimedes in the Middle Ages,* 1332–41. Clagett (1340) points out that Petrarch went beyond his medieval predecessors in seeking out the Latin sources on Archimedes but that he failed to mention a single work by him; he also notes that William of Moerbeke's translation of Archimedes from ca. 1269 could have passed through Avignon while Petrarch was in the area.

5. The works from which Petrarch took this material were all on his list of favorite books. See B. L. Ullman "Petrarch's Favorite Books," *TAPA* 54 (1923): 21–38, reprinted in *Studies in the Italian Renaissance,* 2nd ed. (Rome: Edizioni di Storia e Letteratura, 1973), 113–33. W. Milde's "Petrarch's List of Favorite Books" (*Res Publica Litterarum* 2 [1979]: 229–32) suffers from an unfortunate error in copyediting, which has deprived the *De republica* (book 6) and the *Tusculan Disputations* of their rightful positions at the top of the list. For a facsimile of the list, see H. Rüdiger, "Petrarcas Lieblingsbücher," in *Geschichte der Textüberlieferung der antiken und mittelalterlichen Literatur,* vol. 1 (Zurich: Atlantis Verlag, 1961), 529.

6. Foster, *Petrarch,* 2.

7. For more on this practice of Petrarch's, see C. E. Quillen, *Rereading the Renaissance: Petrarch, Augustine, and the Language of Humanism* (Ann Arbor: University of Michigan Press, 1998), 190–216. Quillen shows extensively how the figure of Augustine contributes to Petrarch's self-fashioning in the *Secretum*.

8. My thanks to Stephen Hinds for allowing me an early look at his *MD* article, which has guided my approach to the material in this chapter. See S. Hinds, "Petrarch, Cicero, Vergil: Virtual Community in *Familiaris* 24.4," *MD* 52 (2004): 157–75 (quote from 157); see also Hinds, "Defamiliarizing Latin Literature," *TAPA* 135 (2005): 49–81. H. Baron (*Petrarch's "Secretum": Its Making and Its Meaning* [Cambridge, MA: Medieval Academy of America, 1985], 12) has observed, "In cases in which a classical model for a humanist's phrasing is known, the motivation of the model often throws light on the imitator's motives."

9. Petrarch *Marc.* 45–46. The passage concludes, "to follow up all these matters explicitly would have been tedious" (*hec omnia expressim exequi longum erat*).

10. Petrarch *Marc.* 48: "But though I was passing right through very many events, two, indeed, appeared by no means worthy of silence" (*sed cum plurima pertransirem, duo quidem haudquaquam digna silentio visa sunt*).

11. Petrarch *Marc.* 50: "*Edixit militibus*" ut Livius ait, "*nequis liberum corpus violaret: cetere prede futura.*" There follows a gap in the text, where Martellotti (*De Viris*, ad loc.) suggests that a sentence has fallen out; it probably described continued resistance in Achradina, for the text resumes with Marcellus's response to a second request for mercy. The episode concludes by repeating the language with which it began: *edicto ut diximus nequis liber violaretur* (it having been decreed, as we said, that no free person be harmed).

12. Petrarch *Marc.* 51: *addit Valerius, ubi lectum nescio, edixisse eum nominatim ut Archimedis capiti parceretur, quanquam illius ingenio atque opera multis et novis machinis excogitatis ad tutelam patrie diu Romanorum victoria retardata esset.*

13. Val. Max. 8.7.7; Firm. Mat. *Math.* 6.30.26; Livy 25.31.10; Cic. *Tusc.* 5.64.

14. Petrarch *Marc.* 56; Livy 25.31.11.

15. In a letter of January 1354 (*Familiares* 18.2), thanking Nicholas Sygeros for a manuscript of Homer, Petrarch says that Homer is "dumb" to him, or, rather, that he is "deaf" to Homer (Wilkins, *Life of Petrarch*, 136). On Petrarch's limited acquaintance with Plutarch, see P. Nolhac, *Pétrarque et l'Humanisme*, 2 vols. (Paris: H. Champion, 1907), 2:122, 128–29.

16. The edict appears at Petrarch *Marc.* 49: "*edixit militibus*" ut Livius ait "*nequis liberum corpus violaret*"; 50: *edicto ut diximus ne quis liber violaretur*; 51: *edixisse eum nominatim ut Archimedis capiti parceretur*.

17. Petrarch *Marc.* 52: *Fuit hic vir quidem insignis astrologus, etsi eum Iulius Firmicius et ipse siculus, nescio an invidia que inter pares precipue ac vicinos regnat an quia sic opinaretur, mechanicum summum dicat, cum untrunque vere et astrologus ingens et mechanicus fuerit, repertorque et fabricator egregius operum diversorum.* Petrarch is following classical usage in calling an astronomer an *astrologus*. See *OLD*, s.v. *astrologus* 1.

18. Petrarch *Marc.* 47: *Tam multa sunt enim et tam varia, ut mirum non sit si fatigaverint bellatores, cum et scriptores fatigaverint et lectores adhuc hodie fatigent. Quem laborem et michi pariter et lectori demere brevi verborum ambitu consilium fuit.* In its general sentiment and in some of its vocabulary (*fatigare, labor*), this passage echoes Livy's expression of relief at having reached the end of the Second Punic War (31.1.1–2).

19. These factors draw attention to Petrarch's role as a biographer employing the

method that he would (later) describe in the preface (*Prohoemium* 3): "I joined together some things that had been said selectively in other authors, and from the sayings of different writers, made one" (*quedam que apud alios carptim dicta erant coniunxi et ex diversorum dictis unum feci*).

20. Although these parts of the narrative give the impression of being glosses, the facsimile of Petrarch's Livy (Harlean 2493) shows no sign of such marginalia at this point in the text. See G. Billanovich, *La tradizione del testo di Livio e le origini dell' umanismo*, vol. 2, *Il Livio del Petrarca e del Valle* (Padua: Antenore, 1981).

21. Petrarch *RM* 1.23. The passage is quoted in Clagett, *Archimedes in the Middle Ages*, 1338–39.

22. Val. Max. 8.7 ext. 7: '*noli' inquit, 'obsecro, istum disturbare'*; Petrarch *Marc* 53: '*Oro,' ait, 'ne hunc michi pulverem confundas'*; Petrarch *RM* 1.23.7: *ne pulverem sibi suum confunderet obsecrabat*.

23. See especially Yates, *Art of Memory* 1–26. According to Yates (101–4), Petrarch was considered a major figure in memory treatises, although he did not leave one behind. Yates guesses that the *RM* was what earned him his reputation as an authority on memory. On Petrarch's exploration of landscape and on its importance for the growth of historicism, see Thomas M. Greene, *The Light in Troy: Imitation and Discovery in Renaissance Poetry* (New Haven and London: Yale University Press, 1982), 88–93. On wandering as a favorite metaphor of Petrarch's, see G. Mazzotta, *The Worlds of Petrarch* (Durham: Duke University Press, 1993), 19–20.

24. Petrarch *RM* 1.37.18. See Billanovich, *Rerum Memorandarum*, cxxiv–cxxx.

25. Petrarch *RM* 1.11.3: "But although both my respect for the queen of cities and the surpassing glory of her achievements deserve that, chronological order in presenting examples set aside [*posthabita ratione temporum in exemplorum relatione*], the Roman ones always take first place."

26. Cf. Valerius Maximus's arrangement of examples illustrating *industria*, in contrast to which Petrarch's interest in the overlap of time, geographical space, and the space created by the text is all the more striking and approaches even Livy's. On this overlap, see Christina S. Kraus, "No Second Troy: Topoi and Refoundation in Livy, Book V," *TAPA* 124 (1994): 267–89; Jaeger, *Livy's Written Rome*.

27. Petrarch *RM* 1.22.3: *sed iam satis multa de nostris; hinc michi procul ab his litoribus navigandum et in amplissimum studiorum pelagus prora flectenda est.*

28. Petrarch *RM* 1.23.1 The overlap of narrator, story, text, and geography is expressed again in the closing words of the Archimedes passage. See Petrarch *RM* 1.23.9: "These things have been said at, perhaps, greater length than necessary, in memory of him whose story [*historiam*], I do not think, will encounter [*occursuram*] me again in this work."

29. Petrarch *RM* 1.23.1: *libet igitur orationis mee cimbam in portu siracusio parumper alligare, dum salutato Archimede proficisor in Graciam.*

30. Hinds, "Petrarch, Cicero, Vergil," 157–61.

31. See, e.g., Pliny *N.H.* 7.208; *Aen.* 6.303; Hor. *Carm.* 2.3.28.

32. Petrarch *RM* 1.23.2: *sed maria et terras et celum omne percurrens meditatione liberrima, quo penetrare acies humana non poterat oculos mentis intendit.*

33. Petrarch *RM* 1.23.3.

34. Livy 24.34.2; Cic. *Tusc.* 1.63.

35. Petrarch introduces this antithesis in the first half, when he points out Archi-

medes' prominence in discussions of the loftiest inquiries that inquiry in heavenly matters has broadcast on earth.

36. This sense of regret is missing from the first two exempla, Caesar and Augustus, even though we learn of Augustus's censorship and self-censorship: he forbade Caesar's youthful writings to be made public and destroyed his own tragedies and his epic *Sicilia*. Although Petrarch does not criticize destruction that took place in antiquity, that of his own age (*etas nostra RM* 1.16.5, 1.17.2, 1.18.2) is another matter.

37. Petrarch *RM* 1.16.5–7: *bonum opus, nisi quod libertini hominis cura cumulaverat ingenuorum torpor effudisset.*

38. Petrarch *RM* 1.17.2, 1.18.2.

39. *Ego itaque, cui nec dolendi ratio deest nec ignorantie solamen adest, velut in confinio duorum populorum constitutus ac simul ante retroque prospiciens, hanc non acceptam a patribus querelam ad posteros deferre volui.*

40. "We have sailed past life, Lucilius [*praenavigavimus, Lucili, vitam*], and just as the lands and cities fall back [*terraeque urbesque recedunt*] as they do in the sea, according to our Virgil, so too in this course of most swiftly moving time, we have left behind first boyhood, then youth, then whatever it is that is in between maturity and old age, placed on the borders of each [*in utriusque confinio positum*], then the best years of old age itself. Finally the common end of the human race begins to come into view [*novissime incipit ostendi publicus finis generis humani*]." See Billanovich, *Rerum Memorandarum*, ad loc.

41. Petrarch's sensitivity to what Stephen Hinds ("Petrarch, Cicero, Vergil," 160) has called "different layers of pastness" means that he can think about different layers of the past with more or less of a sense of loss.

42. For the *studium* of the ancients, see, for example, Petrarch *RM* 1.15.2 (Cicero), 1.16.1(Tiro), 1.17.1 (Sallust). At 1.19.2, Petrarch fears that volumes "labored over by the enthusiasms and wakefulness of our ancestors" (*maiorum studiis vigiliisque elaboratos*) will be lost.

43. *RM* 1.23.2: *quo penetrare acies humana non poterat oculos mentis intendit.*

44. Hinds ("Petrarch, Cicero, Vergil," 166) points out Petrarch's interest in this kind of synchronicity.

45. Thomas M. Greene ("Resurrecting Rome: The Double Task of the Humanist Imagination," in *Rome in the Renaissance: The City and the Myth*, ed. P. A. Ramsey [Binghamton, NY: SUNY Binghamton Press, 1982], 41–54) points out the importance to the humanists of "the imagery of exhumation, bringing to light, rebirth, restoration, resurrection" (41).

46. Cf. Cic. *Tusc.* 5.64: *saeptum undique et vestitum vepribus et dumetis;* Petrarch *Marc.* 55: *De viris illustribus: vepribus obsitum;* Petrarch *RM* 1.23.8: *inter densissimos vepres.* Cf. also Cic. *Tusc.: ignoratum;* Petrarch *De viris illustribus: ignotum;* Petrarch *RM: incognitam.* Greene ("Resurrecting Rome," 45) points out the same imagery in the *De sui ipsius et multorum ignorantia:* "Petrarch locates truth in 'deep and inaccessible . . . caverns,' their approaches obscured by briars and thorns."

47. The *Africa* was begun around 1338 and provisionally drafted by 1343 (T. G. Bergin and A. S. Wilson, *Petrarch's "Africa"* [New Haven: Yale University Press, 1977], ix–xii). Like the *De viris illustribus* and *RM*, it was inspired by Petrarch's early passion for mid-republican Rome, and like them, it remained incomplete. See Foster, *Petrarch*, 155–56. Philip Hardie discusses this dream of Scipio in "After Rome: Renaissance Epic," in

Roman Epic, ed. A. J. Boyle (London and New York: Routledge 1993), 294–302. According to Hardie, "the question of what achievements are truly valuable and the difficulty of aquiring lasting fame are central themes of the *Africa* itself" (294).

48. On the relationship of Cicero's *Somnium Scipionis,* as preserved by Macrobius, to the *Africa,* see Aldo S. Bernardo, *Petrarch, Scipio, and the "Africa": The Birth of Humanism's Dream* (Baltimore: Johns Hopkins University Press, 1962), 103–26.

49. This translation and the following ones are from Bergin and Wilson, *Petrarch's "Africa."*

50. Petrarch *Africa* 2.455 (2.589, trans. Bergin and Wilson). On the transience of glory and for discusion of Petrarch's self-fashioning in the *Africa,* see Hardie, "After Rome," 294–302, with bibliography.

51. Although it is not within the scope of this project to enter into discussion of the *Secretum,* the same argument is made by "Augustine" to "Francesco" at the end of the dialogue; and "Augustine" quotes these passages of the *Africa* in *Secretum* 3, where he argues that glory is transient.

52. On the embedded nature of inscriptions, see Habinek, *Politics of Latin Literature,* 109–14.

CONCLUSION

1. Authier, "Archimedes," 158–59.

2. Richard Martin, "Seven Sages," 116.

3. Cic. *Tusc.* 1.5. Cicero concludes his survey of Greek achievement with mathematics, then turns to oratory: *at contra oratorem celeriter complexi sumus* (but we, on the other hand, quickly embraced the orator). His vocabulary emphasizes the idea of mathematics standing at cultural boundary: *at nos metiendi ratiocinandique utilitate huius artis terminavimus modum* (but we have set the limit of this art [mathematics] by the practical purpose of measuring and reckoning). On the opening of the *Tusculan Disputations,* see Habinek, "Ideology for an Empire," 58–60.

4. *Pro Cluentio,* 87; Att. 12.15, 13.28; *Acad. Pr.* 2.116; *Nat. D.* 2.88.

5. Lactantius *Div. inst.* 2.5.17–19; Sextus Empiricus *Adversus grammaticos* 115; Petrarch *RM* 23.4. Apulieus (*Pro se magia* 16) defending himself against charges of using mirrors in witchcraft, uses Archimedes as an example of one who learned a great deal from mirrors; Ausonius (*Mosella* 300–304) lists him after Daedalus and Philo of Athens among the seven great architects, all of whom would appreciate the buildings along the river. Archimedes also appears in the works of Augustine (*Util. cred.* 6.13; *P.L.* 42.74).

6. In *Class: A Guide through the American Status System* (New York: Summit Books, 1983), 88, Paul Fussell identifies the features of a modern American living room that "confer the air both of archaism and the un-American so essential for upper-class status." He singles out one for special mention: "[I]t's the tabletop obelisk made of marble or crystal, a sly allusion not to Egypt—there would be no class there—but to Paris. And also to Tiffany, known by the cognoscenti to be the main local outlet for these choice items." The book includes a survey by which readers can identify their own positions within the American class structure by assigning points to items in their living rooms. The obelisk wins nine points—the most of any item listed—toward upper-class status. Old, foreign, and appreciated only by the cognoscenti, Archimedes' mechanical sphere would have won high points in an analogous survey of socially aspiring Roman readers.

7. This is "not, it may be thought, an especially apt comparison," observes E. J. Kenney, "but flattering and above all classical" (*The Classical Text* [Berkeley and Los Angeles: University of California Press, 1974], 80). I owe to Kenney both the reference to the preface and the resurrection metaphor at the end of my paragraph. Falkenburg's preface appears in B. Botfield, ed., *Prefaces to the First Editions of the Greek and Roman Classics and of the Sacred Scriptures* (London: H. G. Bohn, 1861), 573.

8. *Quare si gloriari poterat M. Cicero de invento a se Archimedis sepulcro, quod ipse vepribus undique et dumetis obductum Syracusanis ostenderat, multo tu rectius gloriari poteris, quod poetam omnibus numeris absolutissimum primus mortalium divulgaveris.*

9. Cic. *Tusc.* 5.64. Cf. also Falkenburg's *indicavit ille quidem hominis* <u>acutissimi</u> <u>ignoratum</u> *ab omnibus* <u>civibus</u> <u>monumentum</u>, *ut dici posset, ibi Archimedem, non alibi fuisse sepultum* and Cicero's *ita nobilissima Graeciae* <u>civitas</u> . . . *sui* <u>civis</u> *unius* <u>acutissimi</u> <u>monumentum</u> *ignorasset* (*Tusc.* 5.65).

10. The editor was Papadopoulos-Kerameus, in 1899.

11. Heiberg, "Eine Neue Archimedeshandschrift." The first page (235) is a plate reproducing Heiberg's photograph of one page of the text.

12. J. L. Heiberg, ed., *Archimedes Opera Omnia Cum Commentariis Eutochi,* 3 vols. 2nd ed. Leipzig: Teubner, 1910–15, vol. 1, v.

13. The material in this paragraph is based on Netz, *Works of Archimedes,* 2. But see now Netz and Noel, *Archimedes Codex.* The Walters Art Museum maintains a superbly informative Web site for the palimpsest (http://www.archimedespalimpsest .org/palimpsest_history1.html), with images and updates on the work being done on it.

14. The prayer book also contains some ten pages of the Attic orator Hyperides. See Netz and Noel, *Archimedes Codex,* and the Walters Art Museum Web site cited in n. 11.

BIBLIOGRAPHY

Abry, Josèphe-Henriette. "Manilius et Julius Firmicus Maternus, Deux Astrologues sous L'empirè." In *Imago Antiquitatis: Religiones et Iconographe du Monde Romain; Mélanges offerts à Robert Turcan,* ed. N. Blanc and A. Buisson, 35–45. Paris: De Boccard, 1999.

Achard, M. *Sophocle et Archimède: Pères du roman policier.* Liège: Éditions Dynamo, 1960.

Ahl, Frederick M., M. A. Davis, and Arthur Pomeroy. "Silius Italicus." *ANRW* II.32.4 (1986): 2492–2561.

Albrecht, Michael von. *Silius Italicus: Freiheit und Gebundenheit römischer Epik.* Amsterdam: P. Schippers, 1964.

Anderson, Benedict. *Imagined Communities: Reflections on the Origin and Spread of Nationalism.* 2nd ed. London and New York: Verso, 1991.

André, Jean-Marie. "La Rhétorique dans les Préfaces de Vitruve: Le Statut Culturel de la Science." In *Filologia e Forme Letterarie: Studi Offerti a Francesco Della Corte,* ed. Sandro Boldrini, 3:265–99. Urbino: Quattro Venti, 1987.

Appadurai, A., ed. *The Social Life of Things: Commodities in Cultural Perspective.* Cambridge: Cambridge University Press, 1986.

Astin, A. E. "Sources." *CAH,* 2nd ed. (1989): 8.1–16.

Authier, Michael. "Archimedes: The Scientist's Canon." In *A History of Scientific Thought: Elements of a History of Science,* ed. M. Serres, 124–56. Oxford and Cambridge, MA: Blackwell, 1995. Originally published as *Éléments d'Histoire des Science* (Paris: Bordas, 1995).

Averintsev, S. "From Biography to Hagiography." In France and St. Clair, *Mapping Lives: The Uses of Biography,* 19–36.

Badian, Ernst. "The Early Historians." In *Latin Historians,* ed. T. A. Dorey, 1–38. New York: Basic Books, 1966.

Bal, Mieke. "The Narrating and the Focalising: A Theory of the Agents in Narrative." *Style* 17 (1983): 234–69.

Bal, Mieke. "Notes on Narrative Embedding." *Poetics Today* 2, no. 2 (1981): 41–59.

Bal, Mieke. "The Point of Narratology." *Poetics Today* 11, no. 4 (1990): 727–53.

Baldwin, B. "The Date, Identity, and Career of Vitruvius." *Latomus* 49 (1990): 425–34.

Barchiesi, Alessandro. *Speaking Volumes: Narrative and Intertext in Ovid and Other Latin Poets*, ed. and trans. M. Fox and S. Marchesi. London: Duckworth, 2001.

Baron, H. *Petrarch's "Secretum:" Its Making and Its Meaning*. Cambridge, MA: Medieval Academy of America, 1985.

Barton, T. *Ancient Astrology.* London and New York: Routledge, 1994.

Beard, M. "Cicero and Divination: The Formation of a Latin Discourse." *JRS* 76 (1986): 33–46.

Beck, H. "Den Ruhm nicht teilen wollen: Fabius Pictor und die Anfänge des römischen Nobilitätsdiskurses." In Eigler et al., *Formen römischer Geschichtsschreibung*, 73–92.

Beck, H., and U. Walter, eds. *Die frühen römischen Historiker.* Vol. 1, *Von Fabius Pictor bis Cn. Gellius.* Darmstadt: Wissenschaftliche Buchgesellschaft, 2001.

Bérenger, J. *Récherches sur l'aspect idéologique du Principate.* Basel: F. Reinhardt, 1953.

Berger, H., Jr. "Levels of Discourse in Plato's Dialogues." In *Literature and the Question of Philosophy*, ed. A. J. Cascardi, 77–100. Baltimore: Johns Hopkins University Press, 1987.

Bergin, T. G., and A. S. Wilson, trans. *Petrarch's "Africa."* New Haven: Yale University Press, 1977.

Bernardo, Aldo S. *Petrarch, Scipio, and the "Africa": The Birth of Humanism's Dream.* Baltimore: Johns Hopkins University Press, 1962.

Billanovich, G. *La tradizione del testo di Livio e le origini dell' umanismo.* Vol. 2, *Il Livio del Petrarca e del Valle.* Padua: Antenore, 1981.

Billanovich, G., ed. *Rerum Memorandarum Libri.* Vol. 5 of *Edizione nazionale delle opere di Francesco Petrarca.* Florence: Sansoni, 1943.

Blincoe, M. N. "The Use of the Exemplum in Cicero's Philosophical Works." PhD diss., St. Louis University, 1941.

Bliss, F. R. "Valerius Maximus and His Sources: A Stylistic Approach to the Problem." PhD diss., University of North Carolina at Chapel Hill, 1951.

Bloomer, Martin. *Valerius Maximus and the Rhetoric of the New Nobility.* Chapel Hill: University of North Carolina Press, 1992.

Botfield, B., ed. *Prefaces to the First Editions of the Greek and Roman Classics and of the Sacred Scriptures.* London: H. G. Bohn, 1861.

Bourdieu, P. *Distinction: A Social Critique of the Judgement of Taste.* Trans. R. Nice. Cambridge, MA: Harvard University Press, 1984.

Bourdieu, P. *The Field of Cultural Production: Essays on Art and Literature.* Ed. R. Johnson. New York: Columbia University Press, 1993.

Bowen, A. C. "The Art of the Commander and the Emergence of Predicative Astronomy." In Tuplin and Rihll, *Science and Mathematics*, 76–111.

Bradshaw, Gillian. *The Sand-Reckoner.* New York: Forge, 2000.

Bram, Jean Rhys, trans. *Ancient Astrology: Theory and Practice = Matheseos Libri VIII.* Park Ridge, NJ: Noyes, 1975.

Braund, S. M., and C. Gill, eds. *The Passions in Roman Thought and Literature.* Cambridge: Cambridge University Press, 1997.

Brinton, A. "Cicero's Use of Historical Examples in Moral Argument." *Philosophy and Rhetoric* 21, no. 3 (1988): 169–84.

Briscoe, J. "The Second Punic War." *CAH*, 2nd ed. (1989): 8.44–80.

Briscoe, J., ed. *Valerius Maximus: Factorum et Dictorum Memorabilium Libri IX*. Vol. 2. Stuttgart and Leipzig: Teubner, 1998.

Brown, A. "Homeric Talents and the Ethics of Exchange." *JHS* 118 (1998): 165–72.

Brunt, P. A. "Patronage and Politics in the *Verrines*." *Chiron* 10 (1980): 273–89.

Büchner, K., ed. *De Re Publica: Kommentar*. Heidelberg: Carl Winter, 1984.

Buechler, F., and A. Riese, eds. *Anthologia Latinae*. Vol. 1.2. Leipzig: Teubner, 1906. Reprint, Amsterdam: Hakkert, 1964.

Bulmer-Thomas, Ivor, ed. and trans. *Greek Mathematical Works*. Vol. 2, *Aristarchus to Pappus of Alexandria*. Cambridge, MA: Harvard University Press, 1941.

Büttner-Wobst, Th., ed. *Polybii, Historiae*. Vol. 2. Rev. Ludvig Dindorf. 1889. Reprint, Leipzig: Teubner, 1985.

Callebat, L. "Rhétorique et architecture dans le De Architectura de Vitruve. In Gros, *Le Project du Vitruve*, 31–46.

Cameron, A. *Claudian: Poetry and Propaganda at the Court of Honorius*. Oxford: Oxford University Press, 1970.

Canter, H. V. "*Digressio* in the Orations of Cicero." *AJP* 52 (1931): 351–61.

Carawan, E. M. "The Tragic History of Marcellus and Livy's Characterization." *CJ* 80 (1984–85): 131–41.

Champion, Craige B. *Cultural Politics in Polybius's "Histories."* Berkeley, Los Angeles, and London: University of California Press, 2004.

Champlin, Edward. *Final Judgments: Duty and Emotion in Roman Wills, 200 B.C.–A.D. 250*. Berkeley and Los Angeles: University of California Press, 1991.

Chaplin, J. D. *Livy's Exemplary History*. Oxford: Oxford University Press, 2000.

Charette, Fr. "High Tech from Ancient Greece." *Nature* 444 (November 2006): 551–52.

Ciancio, S. *La Tomba di Archimede: Un sepolcro con colonnetta alle porte di Acradina*. Rome, 1965.

Clagett, M. *Archimedes in the Middle Ages*. Vol. 3, *The Fate of the Medieval Archimedes*. Philadelphia: American Philosophical Society, 1978.

Clagett, M. *Greek Science in Antiquity*. New York: Abelard and Shuman, 1955. Reprint, Minola, NY: Dover, 2001.

Clarke, Katherine. *Between Geography and History: Hellenistic Constructions of the Roman World*. Oxford: Oxford University Press, 2000.

Coleman, R. "The Dream of Cicero." *PCPS*, n.s., 10 (1964): 1–14.

Conte, G. B. *Latin Literature: A History*. Trans. J. B. Solodow. Rev. D. Fowler and G. W. Most. Baltimore and London: Johns Hopkins University Press, 1994.

Conte, G. B. *The Rhetoric of Imitation: Genre and Poetic Memory in Virgil and Other Latin Poets*. Trans. Ch. Segal. Ithaca: Cornell University Press, 1986.

Corbeill, A. *Nature Embodied: Gesture in Ancient Rome*. Princeton and Oxford: Princeton University Press, 2004.

Cornell, T. J. *The Beginnings of Rome: Italy and Rome from the Bronze Age to the Punic Wars (c. 1000–264 BC)*. London and New York: Routledge, 1995.

Courtney, E., ed. *The Fragmentary Latin Poets*. Oxford: Oxford University Press, 1993.

Crawford, M. H. *Roman Republican Coinage*. Cambridge: Cambridge University Press, 1974.

Culham, P. "Plutarch and the Roman Siege of Syracuse: The Primacy of Science over Technology." In *Plutarco e le scienze (Atti del IV convegno plutarcheo, Genova-Bocca di Magra, 22–25 Aprile, 1991)*, ed. I. Gallo, 179–97. Genoa: Sage, 1992.

Cuomo, S. *Ancient Mathematics.* London and New York: Routledge, 2001.

Cuomo, S. *Pappas of Alexandria.* Cambridge: Cambridge University Press, 2000.

Davidson, J. "The Gaze in Polybius' *Histories.*" *JRS* 81 (1991): 10–24.

Davies, J. C. "*Reditus ad Rem:* Observations on Cicero's Use of *Digressio.*" *Rheinisches Museum* 131 (1988): 305–15.

Davis, H. H. "Epitaphs and the Memory." *CJ* 53 (1958): 169–76.

De Lacy, P. "Biography and Tragedy in Plutarch." *AJP* 73 (1952): 159–71.

Delz, J., ed. *Silicus Italicus, Punica.* Stuttgart: Teubner, 1987.

Demoen, K. "A Paradigm for the Analysis of Paradigms: The Rhetorical Exemplum in Ancient and Imperial Greek Theory." *Rhetorica* 15 (1997): 125–58.

Dench, Emma. "Beyond Greeks and Barbarians: Italy and Sicily in the Hellenistic Age." In *A Companion to the Hellenistic World,* ed. A. Erskine, 294–310. Malden, MA, and Oxford: Blackwell, 2003.

Dench, Emma. *From Barbarians to New Men: Greek, Roman, and Modern Perceptions of Peoples of the Central Appenines.* Oxford: Oxford University Press, 1995.

Diels, H., and W. Kranz. *Die Fragmenta der Vorsokratiker.* 6th ed. 3 vols. Zurich: Weidmann, 1951. Reprint, 1990–92.

Dihle, A. "Posidonius' System of Moral Philosophy." *JHS* 93 (1973): 50–57.

Dijksterhuis, E. J. *Archimedes.* Trans. C. Dikshoorn. Copenhagen: E. Muskgaard, 1956. Reprint, Princeton: Princeton University Press, 1987. Originally published in Dutch (I. P. Noordhoff: Groningen, 1938).

Dominik, W. J. "Hannibal at the Gates: Programmatising Rome and *Romanitas* in Silius Italicus' *Punica* 1 and 2." In *Flavian Rome: Culture, Image, Text,* ed. A. J. Boyle and W. J. Dominik, 469–97. Leiden and Boston: Brill, 2003.

Dorey, T. A., ed. *Cicero.* New York: Basic Books, 1965.

Dougan, T. W., and R. M. Henry, eds. *Ciceronis Tusculanae Disputationes.* Vol. 2, *Books III–V.* Cambridge: Cambridge University Press, 1905.

Dougherty, C., and L. Kurke, eds. *Cultural Poetics in Archaic Greece.* Oxford: Oxford University Press, 1998.

Douglas, A. E. "Cicero the Philosopher." In Dorey, *Cicero,* 135–70.

Douglas, A. E., ed. *Cicero, "Tusculan Disputations" II & V, with a Summary of III & IV.* Warminster: Aris and Phillips, 1990.

Douglas, A. E. "Form and Content in the *Tusculan Disputations.*" In Powell, *Cicero the Philosopher,* 197–218.

Drachmann, A. G. *The Mechanical Technology of Greek and Roman Antiquity.* Copenhagen: Munksgaard, 1963.

Duckworth, G. E. *The Nature of Roman Comedy: A Study in Popular Entertainment.* 2nd ed. Norman: University of Oklahoma Press, 1994.

Duff, T. *Plutarch's Lives: Exploring Virtue and Vice.* Oxford: Oxford University Press, 1999.

Dugan, J. "How to Make (and Break) a Cicero: Epideixis, Textuality, and Self-Fashioning in the *Pro Archia* and *In Pisonem.*" *CA* 20 (2001): 35–78.

Dugan, J. *Making a New Man: Ciceronian Self-Fashioning in the Rhetorical Works.* Oxford: Oxford University Press, 2005.

Dunbabin, T. J. *The Western Greeks: The History of Sicily and South Italy from the Foundation of the Greek Colonies to 480 B.C.* Oxford: Clarendon, 1948.

Dunkle, J. R. "The Rhetorical Tyrant in Greek and Roman Historiography." *CW* 65 (1971): 12–20.

Dyck, A. "Cicero the Dramaturge." In *Qui Miscuit Utile Dulci: Festschrift Essays for P. Mackendrick,* ed. G. Schmeling and Jon O. Mikalson, 151–64. Wauconda, IL: Bolchazy-Carducci, 1998.

Eckstein, A. M. *Senate and General: Individual Decision-Making and Roman Foreign Relations, 264–194 B.C.* Berkeley, Los Angeles, and London: University of California Press, 1987.

Edel, L., and L. H. Powers, eds. *The Complete Notebooks of Henry James.* Oxford: Oxford University Press, 1987.

Edmunds, Lowell. *Intertextuality and the Reading of Roman Poetry.* Baltimore: Johns Hopkins University Press, 2001.

Edwards, C. *The Politics of Immorality in Ancient Rome.* Cambridge and New York: Cambridge University Press, 1993.

Eigler, U., U. Gotter, N. Luraghi, and U. Walter, eds. *Formen römischer Geschichtsschreibung von den Anfängen bis Livius.* Darmstadt: Wissenschaftliche Buchgesellschaft, 2003.

Einarson, B., and P. H. De Lacy, trans. *Plutarch, "Moralia."* Vol. 14. Cambridge, MA: Harvard University Press, 1967.

Erffa, Helmut von, and Allen Staley. *The Paintings of Benjamin West.* New Haven: Yale University Press, 1986.

Errington, R. M. "Rome and Greece to 205 B.C." *CAH,* 2nd ed. (1989): 8.81–106.

Erskine, A. "Cicero and the Expressions of Grief." In Braund and Gill, *The Passions in Roman Thought,* 36–47.

Fantham, E. *Comparative Studies in Republican Latin Imagery.* Toronto: University of Toronto Press, 1972.

Fargues, P. *Claudian: Études sur sa poésie et son temps.* Paris: Hachette, 1933.

Feeney, Dennis. "The Beginnings of a Literature in Latin." *JRS* 95 (2005): 226–40.

Feldherr, Andrew. "Cicero and the 'Invention' of Roman History." In Eigler et al., *Formen römischer Geschichtsschreibung,* 196–212.

Feldherr, Andrew. *Spectacle and Society in Livy's History.* Berkeley and Los Angeles: University of California Press, 1998.

Finley, M. I. *Economy and Society in Ancient Greece.* Ed. B. D. Shaw and R. P. Saller. New York: Viking, 1981.

Fleury, A., ed. and trans. *Vitruve.* Vol. 1. Paris: Les Belles Lettres, 1990.

Flower, Harriet. *Ancestor Masks and Aristocratic Power in Roman Culture.* Oxford: Oxford University Press, 1996.

Flower, Harriet. "'Memories' of Marcellus: History and Memory in Roman Republican Culture." In Eigler et al., *Formen römischer Geschichtsschreibung,* 39–52.

Flower, Harriet. "The Tradition of the *Spolia Opima.*" *CA* 19, (2000): 34–64.

Foster, Kenelm. *Petrarch: Poet and Humanist.* Edinburgh: Edinburgh University Press, 1984.

Fowler, Don P. "First Thoughts on Closure: Problems and Prospects." *MD* 22 (1989): 75–122.

Fowler, Don P. *Roman Constructions.* Oxford: Oxford University Press, 2000.

Fox, M. "Dialogue and Irony in Cicero." In Sharrock and Morales, *Intratextuality,* 263–86.

Fox, M. *Roman Historical Myths.* Oxford: Oxford University Press, 1996.

France, Peter, and William St. Clair, eds. *Mapping Lives: The Uses of Biography.* Oxford and New York: Oxford University Press, 2002.

Freeth, T., Y. Bitsakis, X. Moussas, J. H. Seiradiakis, A. Tselikas, H. Mangou, M. Zafeiro-poulou, et al. "Decoding the Ancient Greek Astronomical Calculator Known as the Antikythera Mechanism." *Nature* 444 (November 2006): 587–91.

Frier, B. W. *Libri Annales Pontificum Maximorum: The Origins of the Annalistic Tradition.* Papers and Monographs of the American Academy in Rome 27. Rome: American Academy in Rome, 1979. Reprinted with a new introduction. Ann Arbor: University of Michigan Press, 1999.

Fuhrmann, M. *Cicero and the Roman Republic.* Trans. W. E. Yuill. Oxford and Cambridge, MA: Blackwell, 1992.

Fussell, Paul. *Class: A Guide through the American Status System.* New York: Summit Books, 1983.

Galinsky, Karl. *Augustan Culture: An Interpretive Introduction.* Princeton: Princeton University Press, 1996.

Gallagher, Robert. "Metaphor in Cicero's *De Re Publica*." *CQ* n.s. 51, no. 2 (2001): 509–19.

Galluzzi, P. *Archimede e la Storia delle Matematiche nella Galleria degli Uffizi.* Syracuse: A. Lombardi, 1989.

Gee, Emma. "Cicero's Astronomy." *CQ* n.s. 51, no. 2 (2001): 520–36.

Gee, Emma. *Ovid, Aratus, and Augustus: Astronomy in Ovid's "Fasti."* Cambridge: Cambridge University Press, 2000.

Gee, Emma. "*Parva Figura Poli:* Ovid's *Vestalia* (*Fasti* 6.249–468) and the *Phaenomena* of Aratus." *PCPS* n.s., 43 (1997): 21–39.

Geiger, J. "Contemporary Politics in Cicero's *De Republica*." *CP* 79 (1984): 38–43.

Geiger, J. "Plutarch's Parallel Lives: The Choice of Heroes." *Hermes* 109 (1981): 85–104. Reprinted in Scardigli, *Essays on Plutarch's Lives*, 165–90.

Gelzer, M. *Cicero: Ein Biographischer Versuch.* Wiesbaden: F. Steiner, 1969.

Genette, Gerard. *Narrative Discourse: An Essay in Method.* Trans. J. E. Lewin. Ithaca: Cornell University Press, 1980.

Georgiadou, A. "Bias and Character-Portrayal in Plutarch's Lives of Pelopidas and Marcellus." *ANRW* II.33.6 (1992): 4222–57.

Georgiadou, A. "The Corruption of Geometry and the Problem of Two Mean Proportionals." In *Plutarco e le scienze (Atti del IV convegno plutarcheo, Genova-Bocca di Magra, 22–25 Aprile, 1991),* ed. I. Gallo, 147–64. Genoa: Sage, 1992.

Gildenhard, Ingo. "The 'Annalist' before the Annalists: Ennius and His *Annales*." In Eigler et al., *Formen römischer Geschichtsschreibung,* 93–114.

Ginsburg, Judith. *Representing Agrippina: Constructions of Female Power in the Early Roman Empire.* Oxford: Oxford University Press, 2006.

Gowing, Alain. *Empire and Memory: The Representation of the Roman Republic in Imperial Culture.* Cambridge: Cambridge University Press, 2005.

Graff, J. *Ciceros Selbstauffassung.* Heidelberg: Carl Winter, 1963.

Graver, M. *Cicero on the Emotions: "Tusculan Disputations" 3 and 4.* Chicago and London: University of Chicago Press, 2002.

Greene, Thomas M. *The Light in Troy: Imitation and Discovery in Renaissance Poetry.* New Haven and London: Yale University Press, 1982.

Greene, Thomas M. "Resurrecting Rome: The Double Task of the Humanist Imagination." In *Rome in the Renaissance: The City and the Myth,* ed. P. A. Ramsey, 41–54. Medieval and Renaissance Texts and Studies 18. Binghamton, NY: SUNY Binghamton Press, 1982.

Gregory, C. A. "Exchange and Reciprocity." In *The Companion Encyclopedia of Anthropology*, ed. Tim Ingold, 911–39. London and New York: Routledge, 1994.

Gros, P., ed. *Le Project du Vitruve: Object, Destinataires et Réception du De Architectura*. Rome: L'École Française, 1994.

Gros, P. "Les statues de Syracuse et les 'dieux' de Tarent." *REL* 57 (1979): 85–114.

Gruen, Erich S. *Culture and National Identity in Republican Rome*. Ithaca: Cornell University Press, 1992.

Gruen, Erich S. *Studies in Greek Culture and Roman Policy*. Leiden and New York: Brill, 1990.

Habicht, C. *Cicero the Politician*. Baltimore: Johns Hopkins University Press, 1990.

Habinek, Thomas N. "Ideology for an Empire in the Prefaces to Cicero's Dialogues." *Ramus* 23, nos. 1–2 (1994): 55–67.

Habinek, Thomas N. *The Politics of Latin Literature*. Princeton: Princeton University Press, 1998.

Habinek, Thomas N. "Science and Tradition in *Aeneid* 6." *HSCP* 92 (1989): 223–55.

Hall, J. B. ed. *Claudianus, Carmina*. Leipzig: Teubner, 1985.

Hallett, Christopher. *The Roman Nude: Heroic Portrait Statuary, 200 BC–AD 300*. Oxford: Oxford University Press, 2005.

Hardie, Philip. "After Rome: Renaissance Epic." In *Roman Epic*, ed. A. J. Boyle, 294–313. London and New York: Routledge, 1993.

Hardie, Philip. "Closure in Latin Epic." In Roberts, Dunn, and Fowler, *Classical Closure*, 139–62.

Hardie, Philip. *The Epic Successors of Virgil*. Cambridge: Cambridge University Press, 1993.

Harrill, J. A. "The Dramatic Function of the Running Slave Rhoda (Acts 12.13–16): A Piece of Greco-Roman Comedy." *New Testament Studies* 46 (2000): 150–57.

Harrison, Th. "Sicily in the Athenian Imagination: Thucydides and the Persian Wars." In Smith and Serrati, *Sicily from Aeneas to Augustus*, 84–96.

Haury, A. "Cicéron et l'astronomie: À propos de *Rep.* 1.22." *REL* 42 (1964): 198–212.

Heath, T. L., ed. *A History of Greek Mathematics*. Oxford: Oxford University Press, 1921.

Heath, T. L., ed. *The Works of Archimedes*. Cambridge: Cambridge University Press, 1897. Supplement, 1912. Reprint, Mineola, NY: Dover, 2002.

Heiberg, J. L., ed. *Archimedes Opera Omnia cum Commentariis Eutocii*. 3 vols. Leipzig: Teubner, 1880–81. 2nd ed., 1910–15.

Heiberg, J. L. "Eine Neue Archimedeshandschrift." *Hermes* 42 (1907): 235–303.

Heiberg, J. L. *Quaestiones Archimedeae*. Copenhagen: Hauniae, 1879.

Heine, O., and M. Pohlenze, eds. *Ciceronis Tusculorum Disputationum, Libri V*. Stuttgart: Teubner, 1957.

Heinze, R. "*Auctoritas*." *Hermes* 60 (1925): 348–66.

Henderson, J. "The Life and Soul of the Party." In Sharrock and Morales, *Intratextuality*, 287–324.

Hinds, S. *Allusion and Intertext: Dynamics of Appropriation in Roman Poetry*. Cambridge: Cambridge University Press, 1998.

Hinds, S. "Defamiliarizing Latin Literature." *TAPA* 135 (2005): 49–81.

Hinds, S. "Petrarch, Cicero, Vergil: Virtual Community in *Familiaris* 24.4." *MD* 52 (2004): 157–75.

Hölkeskamp, K.-J. "*Exempla* und *mos maiorum*: Überlegungen zum kollektiven

Gedächtnis der Nobilität." In *Vergangenheit und Lebenswelt: Soziale Kommunikation, Traditionsbildung und historisches Bewusstsein,* ed. H.-J. Gehrke and A. Möller, 301–38. Tübingen: Gunter Narr Verlag, 1996.

Hopkins, Keith. *Death and Renewal.* Cambridge: Cambridge University Press, 1983.

Hultsch, F. "Archimedes." *RE* 2.1 (1895): 507–39.

Hultsch, F. *Pappi Alexandrini Collectionis Quae Supersunt.* 3 vols. Berlin: Weidmann, 1876–78.

Hunt, H. A. K. *The Humanism of Cicero.* Carlton: Melbourne University Press, 1954.

Jaeger, Mary. "Cicero and Archimedes' Tomb." *JRS* 92 (2002): 49–61.

Jaeger, Mary. "Livy and the Fall of Syracuse." In Eigler et al., *Formen römischer Geschichtsschreibung,* 213–34.

Jaeger, Mary. *Livy's Written Rome.* Ann Arbor: University of Michigan Press, 1997.

James, Sharon L. "Establishing Rome with the Sword: *Condere* in the *Aeneid.*" *AJP* 116 (1995): 623–37.

Janson, T. *Latin Prose Prefaces.* Stockholm: Almqvist and Wiksell, 1964.

Jones, C. P. *Plutarch and Rome.* Oxford: Oxford University Press, 1971.

Kahn, Charles H. *Pythagoras and the Pythagoreans: A Brief History.* Indianapolis: Hackett, 2001.

Kassel, R. *"Quod Versu Dicere Non Est."* *ZPE* 19 (1975): 211–18.

Kenney, E. J. *The Classical Text.* Berkeley and Los Angeles: University of California Press, 1974.

Keyser, Paul. "Archimedes and the Pseudo-Euclidean *Catoptrics:* Early Stages in the Ancient Geometric Theory of Mirrors." *Archives Internationales d'Histoire des Sciences* 35 (1985): 114–15; (1986): 28–105.

Keyser, Paul. "Cicero on Optics (*Att.* 2.3.2)." *Phoenix* 47 (1993): 67–69.

Keyser, Paul. "Orreries, the Date of [Plato] *Letter* ii, and Eudorus of Alexandria." *Archiv für Geschichte der Philosophie* 80 (1998): 241–67.

Keyser, Paul, and Georgia Irby-Massie. "Science, Medicine, and Technology." In *The Cambridge Companion to the Hellenistic World,* ed. G. R. Bugh, 241–64. Cambridge: Cambridge University Press, 2006.

Kidd, D. A. "Notes on Aratus' *Phaenomena.*" *CQ* n.s., 31 no. 2 (1981): 355–62.

Kiessling, Th., ed. *Ioannis Tzetzae Historiarum Variarum Chiliades.* Hildesheim: Georg Olms, 1963.

Konstan, David. "Narrative and Ideology in Livy: Book I." *CA* 5 (1986): 198–215.

Knorr, W. R. "The Geometry of Burning Mirrors in Antiquity." *ISIS* 74 (1983): 53–73.

Kraus, Christina S. "No Second Troy: Topoi and Refoundation in Livy, Book V." *TAPA* 124 (1994): 267–89.

Kraye, J., ed. *The Cambridge Companion to Renaissance Humanism.* Cambridge: Cambridge University Press, 1994.

Kroll, W., F. Skutsch, eds. *Firmicus Maternus, Mathesis.* Vol. 1. Leipzig: Teubner, 1897.

Kroll, W., F. Skutsch, and K. Ziegler, eds. *Firmicus Maternus, Mathesis.* Vol. 2. Stuttgart: Teubner, 1968.

Kurke, Leslie. *Coins, Bodies, Games, and Gold: The Politics of Meaning in Archaic Greece.* Princeton: Princeton University Press, 1999.

Laird, Andrew. *Powers of Expression, Expressions of Power: Speech Presentation and Latin Literature.* Oxford: Oxford University Press, 1999.

Laird, W. R. "Archimedes among the Humanists." *Isis* 82 (1991): 629–38.

Landels, J. G. *Engineering in the Ancient World.* Berkeley and Los Angeles: University of California Press, 1978.

Landels, J. G. "Ship-Shape and Sambuca-Fashion." *JHS* 86 (1966): 69–77.

Latour, Bruno. *We Have Never Been Modern.* Trans. Catherine Porter. Cambridge, MA: Harvard University Press, 1993. Originally published as *Nous n'avons jamais été modernes: Essais d'anthropologie symmétrique* (Paris: La Découvert, 1991).

Laughton, B. *The Drawings of Daumier and Millet.* New Haven: Yale University Press, 1991.

Laughton, Eric. *The Participle in Cicero.* Oxford: Oxford University Press, 1964.

Lawrence, A. W. "Archimedes and the Design of Euryalus Fort." *JHS* 66 (1946): 99–107.

Leeman, A. D. *Orationis Ratio: The Stylistic Theories and Practice of the Roman Orators, Historians, and Philosophers.* 2 vols. Amsterdam: Hakkert, 1963.

Lendon, J. E. *Empire of Honour: The Art of Government in the Roman World.* Oxford: Oxford University Press, 1997.

Leone, P.A.M., ed. *Ioannis Tzetzae, Historiae.* Naples: Università degli Studi di Napoli, 1968.

Levene, D. S. *Religion in Livy.* Leiden: Brill, 1993.

Levy, C. "Cicero and the *Timaeus.*" In *Plato's "Timaeus" as a Cultural Icon,* ed. G. Reydams-Schils, 95–110. Notre Dame: University of Notre Dame Press, 2003.

Libri, G. *Histoire des Sciences mathématiques en Italie, depuis la Renaissance des Lettres jusqu'à la fin du dix-septième siècle.* 4 vols. Paris: J. Renouardi, 1838–41.

Lintott, A. W. "Imperial Expansion and Moral Decline in the Roman Republic." *Historia* 21 (1972): 626–38.

Litchfield, H. W. "National *Exempla Virtutis* in Roman Literature." *HSCP* 24 (1914): 1–71.

Lomas, K. "Between Greece and Italy: An External Perspective on Culture in Roman Sicily." In Smith and Serrati, *Sicily from Aeneas to Augustus,* 161–73.

MacKendrick, P., and K. L. Singh. *The Philosophical Books of Cicero.* New York: St. Martin's Press, 1989.

MacMullen, R. "Hellenizing the Romans (2nd Century B.C.)." *Historia* 40 (1991): 419–38.

Manetti, G. "The Concept of the Sign from Ancient to Modern Semiotics." In *Knowledge through Signs: Ancient Semiotic Theories and Practices,* ed. G. Manetti, 11–40. Bologna: Brepols, 1996.

Manetti, G. *Theories of the Sign in Classical Antiquity.* Trans. Christine Richardson. Bloomington and Indianapolis: Indiana University Press, 1993.

Mann, Nicholas. "The Origins of Humanism." In Kraye, *Cambridge Companion to Renaissance Humanism,* 1–19.

Mann, Nicholas. *Petrarch.* Oxford and New York: Oxford University Press, 1984.

Marchant, J. "In Search of Lost Time." *Nature* 444 (November 2006): 534–38.

Marincola, John. *Authority and Tradition in Ancient Historiography.* Cambridge: Cambridge University Press, 1997.

Marincola, John. "Marcellus at Syracuse (Livy xxv.24.11–15): A Historian Reflects." In *Studies in Latin Literature and Roman History,* ed. C. Deroux, 12:219–29. Collection Latomus 287. Brussels: Latomus, 2005.

Martellotti, G., ed. *De Viris Illustribus.* Vol. 2 of *Edizione nazionale delle opere di Francesco Petrarca.* Florence: Sansoni, 1964.

Martin, Richard P. "The Seven Sages as Performers of Wisdom." In Dougherty and Kurke, *Cultural Poetics in Archaic Greece,* 108–28. Oxford: Oxford University Press, 1998.

Maslakov, G. "Valerius Maximus and Roman Historiography." *ANRW* II.32.4 (1986): 437–96.

Masterson, M. "Status, Pay, and Pleasure in the *De Architectura* of Vitruvius." *AJP* 125 (2004): 387–416.

Mauss, M. *The Gift: Forms and Functions of Exchange in Archaic Societies.* Trans. Ian Cunnison. Glencoe, IL: Free Press, 1954.

Mayer, R. G. "Roman Historical Exempla in Seneca." In *Sénèque et la prose latine,* ed. O. Reverdin and B. Grange, 141–69. Entretiens Hardt 36. Geneva: Fondation Hardt, 1991.

Mazzotta, G. *The Worlds of Petrarch.* Durham: Duke University Press, 1993.

Mazzuchelli, G. M. *Notizie Istoriche e Critiche intorno alla Vita, alla Invenzioni ed agli Scritti de Archimede Siracusano.* Brescia: G. M. Rizzardi, 1737.

McDonnell, M. "Roman Aesthetics and the Spoils of Syracuse." In *Representing War in Ancient Rome,* ed. S. Dillon and K. E. Welsch, 68–90. Cambridge: Cambridge University Press, 2006.

McDonnell, M. *Roman Manliness: Virtus and the Roman Republic.* Cambridge: Cambridge University Press, 2006.

McEwen, Indra. *Vitruvius: Writing the Body of Architecture.* Cambridge, MA: MIT Press, 2003.

McGowan, Elizabeth P. "Tomb Marker and Turning Post: Funerary Columns in the Archaic Period." *AJA* 99, no. 4 (1995): 615–32.

Michel, A. "Rhétorique et Philosophie dans les 'Tusculanes.'" *REL* 39 (1961): 158–71.

Milde, W. "Petrarch's List of Favorite Books." *Res Publica Litterarum* 2 (1979): 229–32.

Miles, G. *Reconstructing Early Rome.* Ithaca: Cornell University Press, 1995.

Miles, M. "Cicero's Presentation of Gaius Verres: A Roman View of the Ethics of Acquisition of Art." *International Journal of Cultural Property* 2 (2002): 28–49.

Mills, A. A., and R. Clift. "Reflections on the 'Burning Mirrors of Archimedes' with a Consideration of the Geometry and Intensity of Sunlight Reflected from Plane Mirrors." *European Journal of Physics* 13 (1992): 268–79.

Momigliano, Arnaldo. *Alien Wisdom: The Limits of Hellenization.* Cambridge: Cambridge University Press, 1975.

Momigliano, Arnaldo. *The Development of Greek Biography.* 2nd ed. Cambridge, MA: Harvard University Press, 1993.

Monat, P., ed. and trans. *Mathesis/Firmicus Maternus.* Paris: Les Belles Lettres, 1992.

Morrow, Glen, trans. *Proclus, "A Commentary on the First Book of Euclid's 'Elements.'"* Princeton: Princeton University Press, 1970.

Mugler, C., ed. and trans. *Archimède, Oeuvres.* Vol. 1. Budé ed. Paris: Les Belles Lettres, 2003.

Münzer, F. "Claudius." *RE* 3.2:2738–55.

Münzer, F. "Fabius Pictor." *RE* 6.2:1836–41.

Murgia, C. "The Date of the Helen Episode." *HSCP* 101 (2003): 405–26.

Murphy, T. "Cicero's First Readers: Epistolary Evidence for the Dissemination of His Works." *CQ* n.s. 48 no. 2 (1998): 492–505.

Nagel, T. *Mortal Questions.* Cambridge and New York: Cambridge University Press, 1979.

Netz, Reviel. "Greek Mathematicians: A Group Picture." In Tuplin and Rihll, *Science and Mathematics,* 196–216.

Netz, Reviel. *Ludic Proof: Greek Mathematics and the Alexandrian Aesthetic.* Cambridge: Cambridge University Press, forthcoming.

Netz, Reviel. "Proof, Amazement, and the Unexpected." *Science* 298 (2002): 967–68.

Netz, Reviel. *The Shaping of Deduction in Greek Mathematics: A Study in Cognitive History.* Cambridge: Cambridge University Press, 1999.

Netz, Reviel. *The Works of Archimedes Translated into English.* Vol. 1, *The Two Books on the Sphere and the Cylinder.* Cambridge: Cambridge University Press, 2004.

Netz, Reviel, and William Noel. *The Archimedes Codex.* London: Weidenfeld and Nicolson, 2007.

Nicolet, C. *Space, Geography, and Politics in the Early Roman Empire.* Ann Arbor: University of Michigan Press, 1991.

Nisbet, R. G. M., and M. Hubbard. *A Commentary on Horace, "Odes," Book I.* Oxford: Oxford University Press, 1970.

Nolhac, P. *Pétrarque et l'Humanisme.* 2 vols. Paris: H. Champion, 1907.

Ogilvie, R. M. *A Commentary on Livy, Books 1–5.* Oxford: Oxford University Press, 1965.

Osborne, C. *Eros Unveiled: Plato and the God of Love.* Oxford: Clarendon, 1994.

Pape, M. *Griechische Kunstwerke aus Kriegsbeute und ihre öffentliche Austellung in Rome.* Hamburg Diss., 1975.

Pease, A. S., ed. *M. Tulli Ciceronis De Natura Deorum Libri III.* 2 vols. Darmstadt: Wissenschaftliche Buchgesellschaft, 1968.

Pelling, C. "Is Death the End? Closure in Plutarch's Lives." In Roberts, Dunn, and Fowler, *Classical Closure,* 228–50.

Pelling, C. *Plutarch and History: Eighteen Studies.* London: Classical Press of Wales; London: Duckworth, 2002.

Perrin, B., trans. *Plutarch's Lives.* 11 vols. Loeb Classical Library. London: W. Heinemann; New York: G. P. Putnam's Sons; Cambridge, MA: Harvard University Press, 1914–67.

Petrain, D. "Gems, Metapoetics, and Value: Greek and Roman Responses to a Third-Century Discourse on Precious Stones." *TAPA* 135, (2005): 329–57.

Pigler, A. *Barockthemen: Eine Auswahl von Verzeichnissen zur Ikonographie des 17. und 18. Jahrhunderts.* 3 vols. Budapest: Akadémiai Kiadó, 1974.

Plass, P. *Wit and the Writing of History.* Madison: University of Wisconsin Press, 1988.

Pohlenz, M. "Cicero *De re publica* als Kunstwerk." In *Festschrift R. Reitzenstein,* ed. Eduard Fraenkel and Hermann F. Fränkel, 70–105. Leipzig and Berlin: Teubner, 1931. Reprinted in *Kleine Schriften* (Hildesheim: G. Olms, 1965), 374–409.

Pollitt, J. J. "The Impact of Greek Art on Rome." *TAPA* 108 (1978): 155–74.

Pomeroy, A. "Silius Italicus as '*Doctus Poeta.*'" *Ramus* 18, nos. 1–2 (1989): 119–39.

Powell, J. G. F., ed. *Cicero the Philosopher: Twelve Papers.* Oxford: Oxford University Press, 1995.

Powell, J. G. F. *M. Tulli Ciceronis De Re Publica, De Legibus, Cato Maior De Senectute, Laelius De Amicitia.* Oxford: Oxford University Press, 2006.

Price, D. *Gears from the Greeks.* Transactions of the American Philosophical Society, n.s., vol. 64, pt. 7. Philadelphia: American Philosophical Society, 1974.

Quacquarelli, Antonio. "La fortuna di Archimede nei retore negli autori cristiani antichi." *Rendiconti del Seminario Matematico di Messina* 1 (1960–61): 10–50. Reprinted in *Saggi patristici—Retorica ed esegesi biblica.* Quaderni di *Vetera Christianorum* 5 (Bari: Adriatica Editrice, 1971), 381–424.

Quillen, C. E. *Rereading the Renaissance: Petrarch, Augustine, and the Language of Humanism.* Ann Arbor: University of Michigan Press, 1998.

Ramsey, J. T. ed. *Cicero, "Philippics" I–II,* Cambridge: Cambridge University Press, 2003.

Rawson, E. *Cicero: A Portrait.* Ithaca: Cornell University Press, 1975.

Rawson, E. "Cicero the Historian and Cicero the Antiquarian." *JRS* 62 (1972): 33–45.

Rawson, E. *Intellectual Life in the Late Roman Republic.* Baltimore: Johns Hopkins University Press, 1985.

Rawson, E. *Roman Culture and Society.* Oxford: Oxford University Press, 1991.

Rawson, E. "Roman Tradition and the Greek World." *CAH,* 2nd ed. (1989): 8.422–76.

Reekmans, T. "Verbal Humor in Plutarch and Suetonius' Lives." *Ancient Society* 23 (1992): 189–232.

Reeve, M. D. "Classical Scholarship." In Kraye, *Cambridge Companion to Renaissance Humanism,* 20–46.

Reynolds, L. D., ed. *Cicero, "De Finibus."* Oxford: Oxford University Press, 1998.

Riccardi, L. A. Review of *Vitruvius: Writing the Body of Architecture,* by I. McEwen. *Aestimatio* 2 (2005): 136–41.

Riggsby, A. M. *Crime and Community in Ciceronian Rome.* Austin: University of Texas Press, 1999.

Ripoll, François. "Silius Italicus et Cicéron." *LEC* 68 (2000): 2–3.

Rives, J. B. "Marcellus and the Syracusans." *CP* 88 (1993): 32–35.

Roberts, D. H., Fr. Dunn, and D. Fowler, eds. *Classical Closure: Reading the End in Greek and Latin Literature.* Princeton: Princeton University Press, 1997.

Roberts, M. J. *The Jeweled Style: Poetry and Poetics in Late Antiquity.* Ithaca: Cornell University Press, 1989.

Rogers, B. S. "Great Expeditions: Livy on Thucydides." *TAPA* 116 (1986): 335–52.

Roller, M. "Exemplarity in Roman Culture: The Cases of Horatius Cocles and Cloelia." *CP* 99, (2004): 1–56.

Rossi, A. "The Tears of Marcellus: History of a Literary Motif in Livy." *Greece and Rome* 47, no. 1 (2000): 56–66.

Ruch, M. "La composition du *de republica." REL* 26 (1948): 157–71.

Ruch, M. "Météorologie, Astronomie, et Astrologie chez Cicéron." *REL* 32 (1954): 200–219.

Rüdiger, H. "Petrarcas Lieblingsbücher." In *Geschichte der Textüberlieferung der antiken und mittelalterlichen Literatur,* 1:526–37. Zurich: Atlantis Verlag, 1961.

Ruju, P. Alessandra Maccioni, and Marco Mostert. *The Life and Times of Guglielmo Libri (1802–1869), Scientist, Patriot, Scholar, Journalist, and Thief: A Nineteenth-Century Story.* Hilversum: Verloren Publishers, 1995.

Saller, R. *Patriarchy, Property, and Death in the Roman Family.* Cambridge: Cambridge University Press, 1994.

Santini, C. *Silius Italicus and His View of the Past.* Amsterdam: J. C. Gieben, 1991.

Scardigli, B., ed. *Essays on Plutarch's Lives.* Oxford: Oxford University Press, 1995.

Schlachter, A., and F. Gisinger. "Der Globus: Seine Entstehung und Verwendung in der Antike." *Stoicheia* 8, Leipzig and Berlin, 1927.

Schmidt, P. L. "Cicero *De Re Publica:* Die Forschung der letzten fünf Dezennien." *ANRW* I.4 (1973): 262–333.

Schmidt, P. L. "Cicero's Place in Roman Philosophy: A Study of His Prefaces." *CJ* 74 (1978–79): 115–27.

Schofield, M. "Cicero for and against Divination." *JRS* 76 (1986): 47–65.

Schwartz, J. "Recreating an Ancient Death Ray," *New York Times,* October 18, 2005.

Serrati, J. "The Coming of the Romans: Sicily from the Fourth to the First Centuries B.C." In Smith and Serrati, *Sicily from Aeneas to Augustus,* 109–14.

Shackleton-Bailey, D. R., ed. *Valerius Maximus, "Factorum et Dictorum Memorabilium Libri IX."* 2 vols. Cambridge, MA: Harvard University Press, 2000.

Shanzer, D. "Two Clocks and a Wedding." *Romanobarbarica* 14 (1996–97): 225–58.

Sharrock, A., and H. Morales, eds. *Intratextuality: Greek and Roman Textual Relations.* Oxford: Oxford University Press, 2000.

Sihler, E. G. *Cicero of Arpinum: A Political and Literary Biography.* New Haven: Yale University Press, 1914.

Simms, D. L. "Archimedes and the Burning Mirrors at Syracuse." *Technology and Culture* 16 (1977): 1–18.

Simms, D. L. "Archimedes and the Invention of Artillery and Gunpowder." *Technology and Culture* 28 (1987): 67–79.

Simms, D. L. "Archimedes in Literature." *BSHM Newsletter* 34 (Summer 1997): 36–37.

Simms, D. L. "Archimedes the Engineer." *History of Technology* 17 (1995): 45–111.

Simms, D. L. "Archimedes' Weapons of War and Leonardo." *British Journal for the History of Science* 21 (1988): 195–210.

Simms, D. L. "Galen on Archimedes: Burning Mirror or Burning Pitch?" *Technology and Culture* 32 (1991): 91–96.

Simms, D. L. "A Problem for Archimedes." *Technology and Culture* 30, (1989): 177–78.

Simms, D. L. "Santa Lucia and Archimedes." *CA News* 13 (1995): 9–10.

Simms, D. L. "The Trail for Archimedes's Tomb." *Journal of the Warburg and Courtauld Institutes* 53 (1990): 281–86.

Simms, D. L., and D. J. Bryden. "Archimedes and the Opticians of London." *Bulletin of the Scientific Instrument Society* 35 (1992): 11–14.

Simms, D. L., and D. J. Bryden. "Archimedes as an Advertising Symbol." *Technology and Culture* 34 (1993): 387–92.

Skidmore, Clive. *Practical Ethics for Roman Gentlemen: The Work of Valerius Maximus.* Exeter: University of Exeter Press, 1996.

Skutsch, O., ed. *The Annals of Q. Ennius.* Oxford: Oxford University Press, 1985.

Small, Jocelyn P. *Wax Tablets of the Mind: Cognitive Studies of Memory and Literacy in Classical Antiquity.* London and New York: Routledge, 1997.

Smith, C., and J. Serrati, eds. *Sicily from Aeneas to Augustus: New Approaches to Archaeology and History.* Edinburgh: Edinburgh University Press, 2000.

Soubiran, J., trans. *Vitruve, De l'architecture, Livre IX.* Paris: Les Belles Lettres, 1969.

Spaltenstein, Fr. *Commentaire des Punica de Silius Italicus.* 2 vols. Geneva: Droz, 1986–90.

Stahl, W. H., and R. Johnson, with E. L. Burge, trans. *Martianus Capella and the Seven Liberal Arts.* Vol. 2, *The Marriage of Philology and Mercury.* New York: Columbia University Press, 1977.

Steel, C. E. W. *Cicero, Rhetoric, and Empire.* Oxford: Oxford University Press, 2001.

Stein, S. *Archimedes: What Did He Do Besides Cry Eureka?* Washington, DC: Mathematical Association of America, 1999.

Stockton, D. *Cicero: A Political Biography.* Oxford: Oxford University Press, 1971.

Swain, S. C. R. "Hellenic Culture and the Roman Heroes of Plutarch." In Scardigli, *Essays on Plutarch's Lives,* 229–64. Originally published in *JHS* 100 (1990): 126–45.

Swain, S. C. R. *Hellenism and Empire: Language, Classicism, and Power in the Greek World, AD 50–250*. Oxford: Oxford University Press, 1996.

Taylor, R. Review of *Vitruvius: Writing the Body of Architecture*, by I. McEwen. *CP* 100 (2005): 284–89.

Trapp, J. B. "Archimedes's Tomb and the Artists: A Postscript." *Journal of the Warburg and Courtauld Institutes* 53 (1990): 286–88.

Treggiari, S. "Home and Forum: Cicero between 'Public' and 'Private.'" *TAPA* 128 (1998): 1–23.

Tuplin, C. J., and T. E. Rihll. *Science and Mathematics in Ancient Greek Culture*. Oxford and New York: Oxford University Press, 2002.

Tzetzes, Ioannis. *Historiarum Variarum Chiliades*. Ed. Theophilus Kiessling. Hildesheim: Georg Olms, 1963.

Ullman, B. L. "Petrarch's Favorite Books." *TAPA* 54 (1923): 21–38. Reprinted in *Studies in the Italian Renaissance*, 2nd ed. (Rome: Edizioni di Storia e Letteratura, 1973), 113–33.

Vasaly, Ann. *Representations: Images of the World in Ciceronian Oratory*. Berkeley and Los Angeles: University of California Press, 1993.

Ver Eecke, Paul. *Les Oeuvres Complètes d'Archimède*. 2nd ed. 2 vols. Paris: A. Blanchard, 1961.

Ver Eecke, Paul. "Note sur une interprétation erronée d'une sentence d'Archimede." *L'Antiquité classique* 24 (1955): 132–33.

Walbank, F. W. *A Historical Commentary on Polybius*. 3 vols. Oxford: Oxford University Press, 1957–77.

Walbank, F. W. *Polybius*. Berkeley and Los Angeles: University of California Press, 1972.

Walker, A. D. "*Enargeia* and the Spectator in Greek Historiography." *TAPA* 123 (1993): 353–77.

Wallace-Hadrill, A. "Greek Knowledge, Roman Power." *CP* 83 (1988): 224–33.

Wardman, A. *Plutarch's Lives*. Berkeley and Los Angeles: University of California Press, 1974.

Wardman, A. "Plutarch's Methods in the Lives." *CQ* n.s. 21 no. 1 (1971): 254–61.

Webb, R. "Picturing the Past: Uses of Ekphrasis in the *Deipnosophistae* and Other Works of the Second Sophistic." In *Athenaeus and His World: Reading Greek Culture in the Roman Empire*, ed. David Braund and John Wilkins, 218–26. Exeter: University of Exeter Press, 2000.

White, K. D. *Greek and Roman Technology*. Ithaca: Cornell University Press, 1984.

White, S. "Cicero and the Therapists." In Powell, *Cicero the Philosopher*, 219–46.

Wilkins, E. H. *Life of Petrarch*. Chicago and London: University of Chicago Press, 1961.

Wills, Jeffrey. *Repetition in Latin Poetry: Figures of Allusion*. Oxford: Oxford University Press, 1996.

Wilson, N. G. *Scholars of Byzantium*. Baltimore: Johns Hopkins University Press, 1983.

Wilson, R. J. A. *Sicily under the Roman Empire: The Archaeology of a Roman Province, 36 BC–AD 535*. Warminster: Aris and Phillips, 1990.

Winter, Franz. *Der Tod des Archimedes*. Berlin: De Gruyter, 1924.

Wiseman, T. P. *Historiography and Imagination: Eight Essays on Roman Culture*. Exeter: University of Exeter Press, 1994.

Wiseman, T. P. *The Myths of Rome*. Exeter: University of Exeter Press, 2004.

Wiseman, T. P. *Remus: A Roman Myth*. Cambridge and New York: Cambridge University Press, 1995.

Wiseman, T. P. *Roman Drama and Roman History.* Exeter: University of Exeter Press, 1998.

Wood, N. *Cicero's Social and Political Thought.* Berkeley and Los Angeles: University of California Press, 1988.

Woolf, Greg. "Monumental Writing and the Expansion of Roman Society in the Early Empire." *JRS* 86 (1996): 22–39.

Wright, J. *Dancing in Chains: The Stylistic Unity of the Comoedia Palliata.* Papers and Monographs of the American Academy in Rome 25. Rome: American Academy in Rome, 1974.

Yates, F. *The Art of Memory.* Chicago: University of Chicago Press, 1966.

Yegül, F. *Baths and Bathing in Classical Antiquity.* New York and Cambridge, MA: MIT Press, 1982.

Zanker, P. *The Mask of Socrates: The Image of the Intellectual in Antiquity.* Trans. A. Shapiro. Berkeley and Los Angeles: University of California Press, 1995.

Zetzel, J. E. G., ed. *Cicero, "De Re Publica," Selections.* Cambridge: Cambridge University Press, 1995.

Zetzel, J. E. G. "Looking Backward: Past and Present in the Late Roman Republic." *Pegasus* 37 (1994): 20–32.

Ziegler, K., ed. *M. Tullius Cicero: De Re Publica.* Leipzig: Teubner, 1958.

Ziegler, K., and H. Gärtner, eds. *Plutarchis Vitae Parallelae.* Vol. 4. Leipzig: Teubner, 1980.

Ziolkowski, A. "*Urbs Direpta,* or How the Romans Sacked Cities." In *War and Society in the Roman World,* ed. J. W. Rich and G. Shipley, 69–91. London and New York: Routledge, 1993: 69–91.

INDEX